Always Already New

Always Already New

Media, History, and the Data of Culture

Lisa Gitelman

The MIT Press
Cambridge, Massachusetts
London, England

First MIT Press paperback edition, 2008

MIT Press books may be purchased at special quantity discounts for business or sales promotional use. For information, please e-mail <special_sales@mitpress.mit.edu>.

This book was set in Perpetua by Graphic Composition, Inc. Printed and bound in the United States of America.

Library of Congress Cataloging-in-Publication Data

Gitelman, Lisa.
Always already new : media, history and the data of culture / Lisa Gitelman.
p. cm.
Includes bibliographical references and index.
ISBN 978-0-262-07271-7 (hc : alk. paper)—978-0-262-57247-7 (pb : alk. paper)
1. Mass media—History. 2. Communication and technology—United States—History.
I. Title.

P90.G4776 2006
302.2309—dc22

2005058066

10 9 8 7 6 5

In memory of
Facundo Montenegro

Contents

Illustrations

Preface

This book started as a very different project than it has ended. At the outset, this was to be a straightforward monograph version of *New Media, 1740–1915* (MIT, 2003), the collection of essays I coedited with Geoffrey B. Pingree. Geoff and the contributors to that volume were teaching me much about doing media history, and I had the additional desire to be both more pointed and more lucid than I had been in *Scripts, Grooves, and Writing Machines* (Stanford, 1999), about the early history of recorded sound. So this is a book that is partly about "when old technologies were new," as Carolyn Marvin so aptly put it. Moreover, it uses the case of recorded sound to open the important question of how media studies might begin to historicize digital media in a sufficiently rigorous way. Both that question and the Marvinesque perspective are here complicated by the suspicion—resident in media studies since at least the 1960s—that media are curiously reflexive as the subjects of history. That is, there is no getting all the way outside or apart from media to "do" history to them; the critic is also always already being "done" by the media she studies.

As I got deeper into this project, what I first assumed was a disjuncture—between old new media and new—turned out also to possess a few crucial elements of continuity. A second level of argument began to emerge—one that explored commonalities between records and documents, if not exactly between phonographs and digital networks or between playing music and retrieving information. To the extent that it has emerged to the foreground, this second-level contention makes this book as much about the humanities as it is about media history. Records and documents are kernels of humanistic thought, of the specifically modern hermeneutical project that has been associated since the nineteenth century with university departments of history and literature as well as many broader, less academic institutions of public memory, like libraries and museums, and other resonant forms of authoritative cultural self-identification, such as anthologies, reference books, bibliographies, and similar compendiums. What these structures all variously entail is the cultural impulse to preserve and interpret, or better yet, to interpret and preserve, since taking their analysis down to the unit level of records and documents helps to reveal the interpretive structures that are always already in play within any urge or act to preserve. Cultures save themselves. And they save themselves according to a host of little-noticed assumptions that are particularly important to stop and think about in the present moment, as saving increasingly becomes a function of today's new media—something that gets done "on" or "to" the hard drive of a server, for instance, and with a digital device.

I am grateful to the friends and colleagues who have supported me in this endeavor with their advice, criticism, patience, warmth, and wisdom. Among them are many who read parts of this project as I was writing it: I am indebted to Jonathan Auerbach, Judy Babbitts, Wendy Bellion, Carolyn Betensky, Gabriella Coleman, Terry Collins, Pat Crain, Ellen Garvey, Katie King, Matt Kirschenbaum, Sarah Leonard, Lisa Lynch, Meredith McGill, Geoff Pingree, Elena Razlogova, Laura Rigal, Alex Russo, Laura Burd Schiavo, and Gayle Wald. I would also like to thank the audiences on whom I tried out so much of this work before the paint was dry, at the Modern Language Association, American Studies Association, and Media in Transition conferences as well as at the Harvard Humanities Center, New School University, University of Iowa, Concordia University, University of Maryland, University of Minnesota, Dibner Institute at MIT, Leslie Center for the Humanities at Dartmouth College, the history department colloquium at Catholic University, and the Center for Cultural Analysis at Rutgers University. My generous hosts at these institutions have included a number of those listed above along with other friends and colleagues who have been of great moral and intellectual support: Jason Camlot, Robert Friedel and Paul Israel, Henry Jenkins and David Thorburn, Robert Levine, Tom Augst, Leah Price and Jonathan Picker, Lauren Rabinovitz, Eric Rothenbuhler, Michael Warner, and Mark Williams. While other colleagues who attended and participated at these many gatherings must here remain nameless, they have been utterly essential to this project. In the same spirit, I thank my colleagues and the students in media studies at Catholic University. You make media studies easy to believe in and fun to do. Finally, special appreciation is due, as always and again, to Claudia, Hillary, and Alix Gitelman.

This book was supported in part by a fellowship from the National Endowment for the Humanities. I was able to take full advantage of the fellowship because of additional generosity on the part of Catholic University. I am grateful to the endowment and the university. Huge thanks as well to Doug Sery, Valerie Geary, Deborah Cantor-Adams, and the staff and readers for The MIT Press for seeing this project through. One portion of chapter 1 revises an essay appearing in *New Media, 1740–1915*. Chapter 2 revises an essay appearing in *Rethinking Media Change: The Aesthetics of Transition* (MIT, 2003), edited by David Thorburn and Henry Jenkins, and contains a brief element drawn from an essay appearing in *Appropriating Technology: Vernacular Science and Social Power* (Minnesota, 2004), edited by Ron Eglash. These essays make new sense together and have been significantly rewritten for the purposes of this book. Any overlap between old and new versions appears by permission.

Always Already New

Introduction: Media as Historical Subjects

This book examines the ways that media—and particularly new media—are experienced and studied as historical subjects. It uses the examples of recorded sound ("new" between 1878 and 1910) and the World Wide Web, since the Web is a core instance or application of what are today familiarly and collectively referred to as "new media." In pairing these examples, I begin with the truism that all media were once new as well as the assumption, widely shared by others, that looking into the novelty years, transitional states, and identity crises of different media stands to tell us much, both about the course of media history and about the broad conditions by which media and communication are and have been shaped.[1] Though presented chronologically in parts I and II, the histories of recorded sound and digital networking rendered here are intended to speak to one other. In particular, I mean to turn "The Case of Phonographs" against "The Question of the Web," and thereby challenge readers to imagine what a meaningful history of today's new media might eventually look like as well as to think about how accounts of media in general should be written.

This, then, is a book about the ways scholars and critics do media history, but it is more importantly about the ways that people experience meaning, how they perceive the world and communicate with each other, and how they distinguish the past and identify culture. Different versions and styles of media history do make a difference. Is the history of media first and foremost the history of technological methods and devices? Or is the history of media better understood as the story of modern ideas of communication? Or is it about modes and habits of perception? Or about political choices and structures? Should we be looking for a sequence of separate "ages" with ruptures, revolutions, or paradigm shifts in between, or should we be seeing more of an evolution? A progress? Different answers to questions like these suggest different intellectual projects, and they have practical ramifications for the ways that media history gets researched and written. Some accounts of media history offer a sequence of inventors and machines, others trace the development of ideas or epistemologies, and still others chart a changing set of social practices, while many combine elements of several such approaches. In each case, history comes freighted with a host of assumptions about what is important and what isn't—about who is significant and who isn't—as well as about the meanings of media, qualities of human communication, and causal mechanisms that account for historical change. If there is a prevailing mode in general circulation today, I think it is a tendency to naturalize or essentialize media—in short, to cede to them a history that is more powerfully theirs than ours.

Naturalizing, essentializing, or ceding agency to media is something that happens at a lexical level every time anyone says "the media" in English, as if media were a unified natural entity, like the wind. This turn of phrase doubtlessly comes about because of widely shared perceptions that today's news and entertainment outlets together comprise a relatively unified institution. So one refers to what "the Media" is doing in the same spirit that one might refer to what "Big Oil" is up to or how the NASDAQ is performing this month. Forget that the word *media* is rightly plural, not singular. Media are. A medium is. And added to the indisputable if thus tacitly granted consolidation of their corporate ownership, there is another reason why the word *media* gets used so vaguely of late. Media are frequently identified as or with technologies, and one of the burdens of modernity seems to be the tendency to essentialize or grant agency to technology. Here is a simple example: when the Hubble Space Telescope was launched in 1990, it was found to have an incorrectly ground mirror, so that it presented a distorted view of space. My daily newspaper reported at the time that the telescope "needs glasses," making a joke of the fact that in effect, the telescope is glasses already. It is a medium. It doesn't squint around on its own except in a metaphoric sense; it mediates between our eyes and the sites of space that it helps us to experience as sights. Other, much less obvious and less cartoony versions of the same confusion tend to crop up in works by media theorists when technology appears as a form of evidence, a matter I shall return to below.

It pays to be careful with language, and yet media seem to be hard to talk or write about with much precision. For that reason, I begin here by working out a broad definition of media before offering an introduction to both the specific case of early recorded sound, and my larger argument about media and doing media history. My purpose is to be as clear as possible in challenging the ways that I think today's new media, in particular, tend casually to be conceived of as what might be called the end of media history. In thus adapting the phrase "the end of history," I adapt the title of an influential article and book by

Francis Fukuyama. Fukuyama proposed what he described as "a coherent and directional History of mankind that will eventually lead the greater part of humanity to liberal democracy" (1992, xii). ("Liberal" in this context means committed to an open, laissez-faire market.) With the cold war over and capitalism ascendant, Fukuyama argued, the end of that History, with a capital H, was more clearly in sight. Whatever the ultimate fate of this thesis—the controversy it sparked was both trenchant and varied—my point is that media, somewhat like Fukuyama's "mankind," tend unthinkingly to be regarded as heading a certain "coherent and directional" way along an inevitable path, a History, toward a specific and not-so-distant end. Today, the imagination of that end point in the United States remains uncritically replete with confidence in liberal democracy, and has been most uniquely characterized by the cheerful expectation that digital media are all converging toward some harmonious combination or global "synergy," if not also toward some perfect reconciliation of "man" and machine. I note cheerfulness because the same view has not always been so happy. Distributed digital networks have long been described as the ultimate medium in another sense: collectively, they are the medium that can survive thermonuclear war.

This overdetermined sense of reaching the end of media history is probably what accounts for the oddly perennial newness of today's new media. It lingers behind the notion that modernism is now "complete" and familiar temporal sensibilities are at an end.[2] And it accounts as well for the many popular histories and documentaries with titles like *The History of the Future, A Brief History of the Future,* and *Inventing the Future.* In scholarship the same sense of ending appears, for instance, in Friedrich A. Kittler's admittedly "mournful" proposition that "the general digitization of channels and information erases the differences among individual media" so that soon, "a total media link on a digital base will erase the very concept of medium" (1999, 1–2).[3] Likewise, according to Peter Lunenfeld, the digital offers "the universal solvent into which all difference of media dissolves into a pulsing stream of bits and bytes" (1999, 7), effectively suggesting "an end to the end-games of the postmodern era" (2000, xxii). By these accounts, media are the disappearing subjects of the very history they motivate.

Let me clarify: all historical subjects are certainly not alike. The histories of science and art, for instance, differ considerably in the construction of their respective subjects. The art historical object from long ago—a vase, painting, or sculpture—is still art today, however much tastes may have changed. But the scientific object from long ago—curing by leeches, the ether, a geocentric solar system, and so on—isn't science at all. It is myth or fiction. Which kind of historical subjects are media? Are they more like nonscientific or scientific objects? The difference between the two is less about the way different kinds of history get written than it is about a deeply held mental map that people share. A legacy of the Enlightenment, this mental map by convention separates human culture from nonhuman nature.[4] Art and other nonscientific pursuits arise from or represent culture, while science represents nature (I am allowing for a lot of play in that word *represents*). All of the modern disciplines are implicated. Some branches of knowledge, like anthropology, highlight the problems of even making the distinction, since the first generations of anthropologists tended to treat culture as if it were nature. Other disciplines, like history itself, illuminate the casual force with which the distinction gets deployed, since the term *history* denotes both the thing we are doing to the past and the past we are doing it to. This linguistic fact of English is equally apparent in the "two uneven but symbolic halves" of every history book. Every history book has an outside introduction, like the one you are reading, as well as an inside or body. In the first, the author explains the plan of her research, and in the second she offers her results, the details of the past at which she has arrived.[5] The combination becomes effective partly to the degree that the split is taken unreflectively by her readers to echo that culture/nature distinction, the outside artfully made and the inside ("just the facts") truthfully, exactingly rendered.

Media muddy the map. Like old art, old media remain meaningful. Think of medieval manuscripts, eight-track tapes, and rotary phones, or semaphores, stereoscopes, and punch-card programming: only antiquarians use them, but they are all recognizable as media. Yet like old science, old media also seem unacceptably unreal. Neither silent film nor black-and-white television seems right anymore, except as a throwback. Like acoustic (nonelectronic) analog recordings, they just don't do the job. The "job" in question is largely though not exclusively one of representation, and a lot of the muddiness of media as historical subjects arises from their entanglement with this swing term. Media are so integral to a sense of what representation itself *is,* and what counts as adequate—and thereby commodifiable—representation, that they share some of the conventional attributes of both art historical objects and scientific ones. Even media that seem less involved with representation than with transmission, like telegraphs, offer keenly persuasive representations of text, space/time, and human presence, in the form of code, connection, and what critics today call "telepresence," that feeling that there's someone else out there on the other end of the line.[6] It is not just that each new medium represents its predecessors, as Marshall McLuhan noted long ago, but rather, as Rick Altman (1984, 121) elaborates, that media represent and delimit representing, so that new media provide new sites for the ongoing and vernacular experience of representation as such.

When I say that this is a book about media as historical subjects, I mean to motivate just this complexity. If *history* is a term that means both what happened in the past and the varied practices of representing that past, then media are historical at several different levels. First, media are themselves denizens of the past. Even the newest new media today come from somewhere, whether that somewhere gets described broadly as a matter of supervening social necessity, or narrowly in reference to some proverbial drawing board and a round or two of beta testing.[7] But media are also historical because they are functionally integral to a sense of pastness. Not only do people regularly learn about the past by means of media representations—books, films, and so on—using media also involves implicit encounters with the past that produced the representations in question. These implicit encounters with the past take many forms. A photograph, for instance, offers a two-dimensional, visual representation of its subject, but it also stands uniquely as evidence, an index, because that photograph was caused in the moment of the past that it represents. Other encounters with the past can be less clear, less causal, and less indexical, as when the viewers of a television newscast are "taken live" to the outside of a building where something happened a little while ago, or when digital images recomplicate the notion of a photographic index altogether.

As my allusion to the Hubble Space Telescope suggests, one helpful way to think of media may be as the scientific instruments of a society at large. Since the late seventeenth century, scientific instruments have emerged as matters of consensus within a community of like-minded and fairly well-to-do people, eventually called scientists. If one scientist or a group of scientists invents a new instrument, they must demonstrate persuasively that the instrument does or means what they say, that it represents the kind and order of phenomena they intend. Other scientists start using the instrument, and ideally, its general acceptance soon helps to make it a transparent fact of scientific practice. Now scientists everywhere use the air pump, say, or the electrophoresis gel without thinking about it. They look through the instrument the way one looks through a telescope, without getting caught up in battles already won over whether and how it does the job. The instrument and all of its supporting protocols (norms about how and where one uses it, but also standards like units of measure) have become self-evident as the result of social processes that attend both laboratory practice and scientific publication.

Media technologies work this way too. Inventing, promoting, and using the first telephones involved lots of self-conscious attention to telephony. But today, people converse through the phone without giving it a moment's thought. The technology and all of its supporting protocols (that you answer "Hello?" and that you pay the company, but also standards like touch-tones and twelve-volt lines) have become self-evident as the result of social processes, including the habits associated with other, related media. Self-evidence or transparency may seem less important to video games, radio programs, or pulp fiction than to telephones, yet as critics have long noted, the success of all media depends at some level on inattention or "blindness" to the media technologies themselves (and all of their supporting protocols) in favor of attention to the phenomena, "the content," that they represent for users' edification or enjoyment.[8] When one uses antique media like stereoscopes, when one encounters unfamiliar protocols, like using a pay telephone abroad, or when media break down, like the Hubble Space Telescope, forgotten questions about whether and how media do the job can bubble to the surface.

When media are new, they offer a look into the different ways that their jobs get constructed as such. Of particular interest in this book are the media that variously do the job of inscription. Like other media, inscriptive media represent, but the representations they entail and circulate are crucially material as well as semiotic. Unlike radio signals, for instance, inscriptions are stable and savable. Inscriptions don't disappear into the air the way that broadcasts do (though radio and television can of course be taped—that is, inscribed). The difference seems obvious, but it is important to note that the stability and savability of inscriptions are qualities that arise socially as well as perceptually. The defining fixity of print as a form of inscription, for example, turns out to have arisen as a social consequence of early modern print circulation as much as from any perceptual or epistemological conditions inherent to printed editions in distinction from manuscript copies. Likewise, the defining scientific or self-evident qualities of landscape photography turn out to have resulted from nineteenth-century practices of illustration and narration as much as from any precision inherent to photographs in distinction from painted panoramas or other forms.[9] The introduction of new media, these instances suggest, is never entirely revolutionary: new media are less points of epistemic rupture than they are socially embedded sites for the ongoing negotiation of meaning as such. Comparing and contrasting new media thus stand to offer a view of negotiability in itself—a view, that is, of the contested relations of force that determine the pathways by which new media may eventually become old hat.

One of the advantages of drawing this analogy between scientific instruments and media is that it helps to locate media at the intersection of authority and amnesia. Just as science enjoys an authority by virtue of its separation from politics and the larger social sphere, media become authoritative as the social processes of their definition and dissemination are separated out or forgotten, and as the social processes of protocol formation and acceptance get ignored.[10] One might even say that a supporting protocol shared by both science and media is the eventual abnegation and invisibility of supporting protocols.[11] Science and media become transparent when scientists and society at large forget many of the norms and standards they are heeding, and then forget that they are heeding norms and standards at all. Yet transparency is always chimerical. As much as people may converse through a telephone and forget the telephone itself, the context of telephoning makes all kinds of difference to the things they say and the way they say them. The same is also true of science: geneticists use *drosophila* (fruit flies) as a kind of instrument, and genetics itself would be substantively different if a different organism were used.[12] The particular authority of science makes this an uncomfortable claim, so crossing over to the other half of the collective mental map renders the point more clearly. Just as it makes no sense to appreciate an artwork without attending to its medium (painted in watercolors or oils? sculpted in granite or Styrofoam?), it makes no sense to think about "content" without attending to the medium that both communicates that content and represents or helps to set the limits of what that content can consist of. Even when the content in question is what has for the last century or so been termed "information," it cannot be considered "free of" or apart from the media that help to define it. However commonplace it is to think of information as separable from, cleanly contained in, or uninformed by media, such thinking merely redoubles a structural amnesia that already pertains.[13]

I define media as socially realized structures of communication, where structures include both technological forms and their associated protocols, and where communication is a cultural practice, a ritualized collocation of different people on the same mental map, sharing or engaged with popular ontologies of representation.[14] As such, media are unique and complicated historical subjects. Their histories must be social and cultural, not the stories of how one technology leads to another, or of isolated geniuses working their magic on the world. Any full accounting will require, as William Uricchio (2003, 24) puts its, "an embrace of multiplicity, complexity and even contradiction if sense is to be made of such" pervasive and dynamic cultural phenomena.

Defining media this way admittedly keeps things muddy. If media include what I am calling protocols, they include a vast clutter of normative rules and default conditions, which gather and adhere like a nebulous array around a technological nucleus. Protocols express a huge variety of social, economic, and material relationships. So telephony includes the salutation "Hello?" (for English speakers, at least), the monthly billing cycle, and the wires and cables that materially connect our phones. E-mail includes all of the elaborately layered technical protocols and interconnected service providers that constitute the Internet, but it also includes both the QWERTY keyboards on which e-mail gets "typed" and the shared sense people have of what the e-mail genre is. Cinema includes everything from the sprocket holes that run along the sides of film to the widely shared sense of being able to wait and see "films" at home on video. Some protocols get imposed, by bodies like the National Institute of Standards and Technology or the International Organization for Standardization. Other protocols get effectively imposed, by corporate giants like Microsoft, because of the market share they enjoy. But there are many others that emerge at the grassroots level. Some seem to arrive *sui generis,* discrete and fully formed, while many, like digital genres, video rentals, and computer keyboards, emerge as complicated engagements among different media. And protocols are far from static. Although they possess extraordinary inertia, norms and standards can and do change, because they are expressive of changeable social, economic, and material relationships.

Nor are technological nuclei as stable as I have just implied. Indeed, much of their coherence as nuclei may be heuristic. (That is, they only look that way when they get looked at.) As Walter Benjamin (1999, 476) noted about historical subjects generally, "The present determines where, in the object from the past, that object's fore-history and after-history diverge so as to circumscribe its nucleus." So it is as much of a mistake to write broadly of "the telephone," "the camera," or "the computer" as it is "the media," and of—now, somehow, "the Internet" and "the Web"—naturalizing or essentializing technologies as if they were unchanging, "immutable objects with given, self-defining properties" around which changes swirl, and to or from which history proceeds.[15] Instead, it is better to specify telephones in 1890 in the rural United States, broadcast telephones in Budapest in the 1920s, or cellular, satellite, corded, and cordless landline telephones in North America at the beginning of the twenty-first century. Specificity is key. Rather than static, blunt, and unchanging technology, every medium involves a "sequence of displacements and obsolescences, part of the delirious operations of modernization," as Jonathan Crary puts it (1999, 13). Consider again how fast digital media are changing today. Media, it should be clear, are very particular sites for very particular, importantly social as well as historically and culturally specific experiences of meaning. For this reason, the primary mode of this book is the case study.

For all of their particularity, media frequently get lumped together by different schools of thought for overarching purposes. If media are sites for experiences of meaning—critics have pondered—to what degree are meaning and its experience determined or circumscribed by technological conditions? To what extent are they imposed or structurally effected by a "culture industry," the combined interests of Hollywood, Bertelsmann, AOL/Time Warner, and an ever dwindling number of multinational media conglomerates? Or are experiences of meaning more rightly produced than determined and imposed? How might production in this case include the ordinary people (who experience meanings) as well as the multinational industry, notwithstanding such a dramatic disparity in their power?[16] This sort of abstract puzzling does have a practical politics. If meanings are imposed by industry, then policing media becomes a viable project: quashing violence on television and labeling offensive lyrics will protect minors from harm and lead to a decrease in violent crime. But if viewers and listeners themselves help variously, literally, to produce the meanings they enjoy, then policing media is pretty much beside the point. Viewers will make of violent content what they will. At stake are two different versions of agency. Either media audiences lack agency or they possess it. Hardly anyone would say the truth can't lie somewhere in between these two extremely reductive positions, but legislators still have to vote either yes or no when the question comes up.

Related questions of agency are vital to media history. As I've already noted, there is a tendency to treat media as the self-acting agents of their own history. Thus, Jay David Bolter and Richard Grusin (1999, 15) write that new media present themselves

> as refashioned and improved versions of other media. Digital media can best be understood through the ways in which they honor, rival, and revise linear-perspective painting, photography, film, television, and print. No medium . . . seems to do its cultural work in isolation from other media, any more than it works in isolation from other social and economic forces. What is new about new media comes from the particular ways in which they refashion older media and the ways in which older media refashion themselves to answer the challenges of new media.

Here, Bolter and Grusin's identification of media as social and economic forces appears amid a lot of syntax that seems to make media into intentional agents, as if media purposefully refashion each other and "do cultural work." However astute their readings of the ways different media compare and contrast at a formal level, Bolter and Grusin have trimmed out any mention of human agents, as if media were naturally the way they are, without authors, designers, engineers, entrepreneurs, programmers, investors, owners, or audiences. Of course Bolter and Grusin know better.[17] People just write this way, Raymond Williams has suggested, because agency is so hard to specify; technological innovation appears autonomous, Williams ([1974] 1992, 129) argues, "only to the extent that we fail to identify and challenge its real agencies."

Ironically, though, critics who do celebrate the real agency of individual inventors sometimes end up a lot like Bolter and Grusin. Kittler's media discourse analysis valorizes Thomas Edison, offering several competing versions of the inventor's agency with regard to the invention of recorded sound. "Edison's phonograph," according to Kittler (1999, 27), "was a by-product of the attempt to optimize telephony and telegraphy by saving expensive copper cables." But Edison also "developed his phonograph in an attempt to improve the processing speed of the Morse telegraph beyond human limitations," Kittler notes, and he did so when "a Willis-type machine [for synthesizing sounds] gave him the idea" and "a Scott-type machine [for drawing sound waves] pushed him towards its realization" (190). Though these statements each sound convincing, complete with human agents and human intentions, Kittler offers no evidence at all to support them. He might have cited from some thousands of pages of existing documentation, from Edison's experimental notebooks or items of correspondence from 1877. Documents from that July, for instance, indicate that Edison was struggling to improve the sibilant articulation of Alexander Graham Bell's telephone. In one technical note from July 18 titled "Speaking Telegraph," Edison (1994, 443–444) comments, "Spkg [speaking] vibrations are indented nicely" on waxed paper by "a diaphram [*sic*] having an embossing point," so that, he reasons, he should be "able to store up & reproduce automatically at any future time the human voice perfectly." This realization could be called the invention of the phonograph, and so could a number of other related actions at Menlo Park, New Jersey, over the next few months. My point is less that Kittler overstates and undercites than that he appears to be arguing backward from what Geoffrey Winthrop-Young and Michael Wutz (1999, xiv) term an "intrinsic technological logic"—a logic Kittler reads as inherent to the phonograph once it was already invented.[18] However extraordinarily rich his sense of media and the "discourse networks" they help to support, it is as if Kittler doesn't need to persuade his readers of details about why or how phonographs were invented because he already knows what phonographs are, and therefore he knows what (and particularly how) they mean. Again, that is to make a medium both evidence and cause of its own history.

In the pages that follow, I have resisted thinking of media themselves as social and economic forces and have resisted the idea of an intrinsic technological logic. Media are more properly the results of social and economic forces, so that any technological logic they possess is only apparently intrinsic. That said, I have also resisted taking a reductively antideterministic position. At certain levels, media are very influential, and their material properties do (literally and figuratively) *matter,* determining some of the local conditions of communication amid the broader circulations that at once express and constitute social relations. This "materiality" of media is one of the things that interests me most.

The advantage of offering finely grained case studies is that it allows these complexities to emerge. I have worked within narrow chronological brackets, both in treating the case of phonographs and that of digital networks, and I have further limited my scope to the cultural geography of the United States, with which I am most familiar. While such a perspective has obvious shortcomings, the detail and specificity of each case permits an account that is exacting, and at the same time broadly suggestive of the ways that new media emerge into and engage their cultural and economic contexts as well as the ways that new media are shaped by and help to shape the semiotic, perceptual, and epistemic conditions that attend and prevail.

By amplifying two specific case studies, one past and one more present, the shape of this book resembles and appreciates the "media archaeologies" produced by a number of recent critics. As Geert Lovink (2003, 11) generalizes the archaeological perspective, "Media archaeology is first and foremost a methodology, a hermeneutic reading of the 'new' against the grain of the past, rather than a telling of the histories of technologies from past to present." By reading digital media "into history, not the other way around," Lovink suggests, the media archaeologist seeks a built-in refusal of teleology, of narrative explanations that smack structurally of the impositions of metahistory.[19] Since telling a story imposes a logic retrospectively onto events, that is, these critics seek to avoid and thereby critique storytelling. (Just as—and at the same time that—no one in cultural studies wants to admit of technological determinism, no one in cultural studies seems to want to be historicist according to any but a "new" historicist paradigm.) This helps to explain Lev Manovich's (2001) "parallels" between Russian constructivist cinema and today's new media. It explains why Alan Liu's (2004b, 72) brilliant comparison of the paper forms used in Taylorist scientific management and today's "encoded discourse" reveals a "surprising bandwidth of connection," in which the past serves only as "an index or placeholder (rather than cause or antecedent) of the future." In short, the impulse to resist historical narrative redraws criticism as a form of "aesthetic" or "literary" undertaking at the same time that it tends to impose a temporal asymmetry.[20] The past is often represented discretely, formally, in isolation—as or by means of anecdote—while the present retains a highly nuanced or lived periodicity, as when Lovink's (2003, 43–44) criticism parses so carefully the mid-1990s' "mythological-libertarian techno-imagination of *Mondo 2000* and *Wired;* the massification of the medium, accompanied by the dotcom craze; [and] the consolidation during the 2000–2002 depression," and the networking of today.[21]

I want to distinguish my method from media archaeology and related cultural studies in several respects. Media archaeology is rightly and productively mindful of historical narrative as a cultural production of the present. The two case studies that follow seek further to pick out related forms of mindfulness in as well as with regard to the past. Why these two cases? Both describe—even, yes, narrate—moments when the future narratability of contemporary events was called into question by widely shared apprehensions of technological and social change as well as by varied engagements—tacit as well as knowing—with what I refer to as "the data of culture": records and documents, the archivable bits or irreducible pieces of modern culture that seem archivable under prevailing and evolving knowledge structures, and that thus suggest, demand, or defy preservation. History in this sense is no less of a cultural production in the past than it is in the present. My first case concerns events that occurred during the extended moment at the end of the nineteenth century when the humanities emerged in something like their present form, both institutionally and epistemologically, becoming what Lawrence Veysey (1999, 52) terms the "special [bulwark] of an orientation toward the past." (The humanities are our past-oriented disciplines: history, English, classics, philosophy, art history, comparative literature.) My second case concerns events that occurred during the extended moment at the end of the twentieth century when the humanities in the United States may have enjoyed the possibility of centralization, in the form of state sponsorship, yet entered what is widely perceived as a period of ongoing "crisis."[22] I offer two case studies in order to benefit from contrast and comparison, not to refine one at the expense of the other. The chronological gap between them has helped me keep "one eye focused on historical variability and the other on [elements of] epistemological constancy" that underwrite the humanities still, and that like all protocols, can be difficult to see without seeking or contriving some penumbra of discontinuity, such as the joint discontinuousness of time frames and newness of new media rendered in these pages.[23]

In chapter 1 I describe the medium of recorded sound as it was first introduced to the U.S. public. During the spring and summer of 1878, audiences could pay to see and hear recordings made and replayed on Edison's initially crude device. A series of lyceum demonstrations across the United States, together with the many newspaper accounts they stimulated, helped to identify the new medium. Then in 1889–1893, audiences got a second look and listen. This time they paid for encounters with an improved version of Edison's machine, adapted to play prerecorded musical selections in public places. Neither endeavor lasted or was profitable for very long. While it is easy to reason in hindsight that these initial endeavors eventually failed because neither the technology nor its supporting protocols had successfully been defined yet, one might also argue that neither the lyceum demonstrations nor the public amusement trade successfully located the U.S.

public that they supposed. Media and their publics coevolve. Because the demonstrations of 1878 have never been studied before in any detail, it has never been clear the extent to which—far from possessing an intrinsic logic of its own—the new medium was experienced as party to the existing, dynamic (and extrinsic) logics of writing, print media, and public speech. Audiences experienced and helped to construct a coincident yet contravening logic for recorded sound, responding to material features of the new medium as well as the contexts of its introduction and ongoing reception and development.

As Jürgen Habermas first proposed and subsequent scholars have elaborated, the extrinsic or cultural logics of print media and public speech are particularly important historically because beginning sometime in the seventeenth century, they doubled as the cultural logic of the bourgeois public sphere. That is, the same assumptions that lay behind the commonsense intelligibility of publication and public speaking *as such* also helped to "determine how the political arena operates," locating an abstract social space for public discussion and opinion, in which some voices, some expressions, were legitimate—and legitimated—while others were constrained.[24] On one level, Edison's phonograph stumbled hard against this public sphere: by intruding on experiences of printedness and public speaking, the phonograph records of 1878 and 1889–1893 abruptly called its commonsense parameters into question, begging a mutual redefinition of print, speech, and public. On another level, however, Edison and his phonographs were themselves part of much larger versions of the same questions already being broached. Though Edison would not, of course, have expressed it this way, he and his invention were part of an ongoing industrialization of communication. (Here's where his telegraphs and telephones fit in too, along with a massive growth and diversification of print media.) The industrialization of communication resulted from as well as abetted new social and economic structures. These new structures served—anything but abruptly—to jeopardize the very commonness and sensibleness of the commonsense intelligibility of publication, and also the boundaries and operations of the political arena. By this account, Edison and the first phonographs didn't stumble against the public sphere as much as they encountered it stumbling. The new medium with its emergent norms and standards at this level actually helped to steady and partly reconstruct a common or normative sense of publicness and an abstract public, one for which recording and playback were intelligible, and for which the logic of phonographs and phonograph records might seem to be intrinsic.

The vague, new "social and economic structures" of the previous paragraph deserve a word of elaboration, since I have described them as causal (if also reciprocal) agents of media history in the nineteenth century. These new social and economic structures included things like modern corporations and the "visible hand" of an emergent managerial class as well as modern markets with centralized trading in securities and commodity futures—familiar characters all, in histories of industrialized communication or "the control revolution," as James Beniger (1986) has called it.[25] Less frequently noted in the same accounts but equally pertinent were concomitant social and economic structures like an emergent class of wage laborers, the emergent demographics of increased immigration and U.S. imperial expansion, and the related emergence of new, urban mass audiences for print media and public spectacle. If the industrialization of communication broadly attended social and economic structures such as these, then the new medium of recorded sound consisted in part of protocols expressive of the relationships they entailed. This is not to suggest that early phonographs were in some respect either managerial or proletarian. Rather, the commonsense intelligibility of the new medium emerged in keeping with a dialectic between control and differentiation, between the traditional public sphere and its potential new constituents. Predictably, the potential new constituents most important to the definition of the new medium were also in some respects the least "other" or alien. Chapter 2 demonstrates in detail that the new medium of recorded sound was deeply defined by women, generally middle-class women, who helped to make it a new, newly intelligible medium for home entertainment.

Chapter 2 follows the new medium out of public places and into private homes. That transit, accomplished with such success around 1895 to 1900, scuttled the expectations of Edison and others who thought of phonographs as business machines for taking dictation. Playback not recording emerged as the primary function of the medium and a commercial bonanza for its corporate owners, although dictation phonographs (Dictaphone was one trade name) would remain continuously available for sale in the United States until the eventual success of magnetic tape recorders after the Second World War. This switch in primary function from dictation to amusement has been popularly explained as both an example of Edison's "accidental genius" (*Wired* 2002, 92) and the inventor Emile Berliner's "killer application" (Naughton 2000, 245), since Berliner envisioned his version of recorded sound, the gramophone, as an amusement device from its first unveiling in 1888.[26] The switch has also been explained as an industrial design triumph: a better power source, cheaper machines, and mass-produced musical recordings. And it has likewise been explained as a culture industry coup: star performers, hit records, major labels, and seductive advertising campaigns. Most accounts agree that consumer demand played a decisive role in making the new medium of recorded sound into a mass medium—one that by 1910 was helping to restructure the ways that Americans experienced music and helping (along with movies, magazines, comics, vaudeville circuits, and the like) to reorient U.S. social life toward ever-increasing leisure consumption.

Consumer demand was decisive, I agree, but part of my argument is that the very categories of consumer and producer are inadequate to explain fully the deep definition of new media. When media are new, when their protocols are still emerging and the social, economic, and material relationships they will eventually express are still in formation, consumption and production can be notably indistinct. The new medium of recorded sound became intelligible as a form of home entertainment according to ongoing constructions of home and public—constructions that relied centrally during the late nineteenth century on changing roles for women, and further, changing experiences of gender and cultural difference. The same broad social contexts have been described as equally, if differently, defining for telephones, monthly magazines, and motion pictures in the same period.[27] Women helped to engender a new mutual logic for media and public life. Protocols and indeed the primary function of the new medium of recorded sound emerged in part according to contexts involving practices as varied as mimicry by vaudevillian comediennes and parlor piano playing by ladylike amateurs, shaped by potently gendered constructions of work and leisure as well as of production and consumption. Even the technical protocols of the medium, like the hardness of recording surfaces and the design of recording styli, emerged partly in response to the timbre of women's voices, which proved tricky to record well (and thus to make public), and therefore informed emerging commonsense norms for A&R (artists and repertory) and emerging commonsense standards of acoustic fidelity.

In short, the definition of new media depends intricately on the whole social context within which production and consumption get defined—and defined as distinct—rather than merely on producers and consumers themselves. This is not to diminish the role of human agents but only to describe more thoroughly where more of them stand in order to resist, as much as possible, the disavowal of underlying economic structures or cultural politics. At the end of the nineteenth century in the United States, the medium of recorded sound helped both to destabilize and to steady or partly reconstruct an abstract sense of publicness, one that increasingly included women, immigrants, and workers—increasingly included "others"—as constitutive members. Of course, rather like Groucho Marx not belonging to any club that would have him as a member, the new sense of public that emerged was different or other than the old, in the least because the new public sphere was increasingly experienced as collective of consumers rather than citizens, increasingly restructured, as Habermas (1989) has indicated, by a cultural premium on publicity and public taste. Not that I wish to romanticize the Habermasian bourgeois public sphere or overstate its debatable explanatory power. The public is a "practical fiction," in Michael Warner's (2002, 8) terms, based in the United States on whiteness and masculinity. Its conception, however, "is unthinkable," Michael Geisler (1999, 99) explains, "without the centripetal power of media to offset the centrifugal force" of social differentiation.[28]

This dialectic between control and differentiation, between existing media publics and their potential new constituents, has emerged in a slightly different form today as a central device in the growing literature on globalization. Intuitively, worldwide digital and satellite communications pull people together, and in doing so they moderate differences and homogenize cultures. In this literature, media serve as instruments of Western cultural imperialism and mature finance capital, creating a global village of increasingly Americanized consumers. Culturally, globalization is a process involving worldwide transfers of technology and translocations of people—migrations, diasporas, and displacements—that is resisted hopelessly, if at all, by the centrifugal pressures of localism. However apposite this dark picture may be, it is painted with a broad brush, the wide strokes of which threaten to blur away the very localism they purport to show in decline and at the same time exaggerate the ways in which today's new media are distinctively new.

It will pay to remember that at the beginning of the twentieth century, the medium of recorded sound formed part of an increasingly global economy marked by flows of capital and commodities on an unprecedented scale—flows that would dwindle abruptly with the First World War and then remain unmatched in magnitude until the end of the century.[29] The new medium depended on a worldwide trade in materials—like German chemicals and Indian lac (the insect secretion required to make the shellac for records)—as well as recording artists, recording studios, and phonograph and gramophone dealers around the world. As Andrew Jones (2001, 54) puts it, "This new (and immensely profitable) industry was—from its very inception—transnational in character." The British Gramophone Company established subsidiaries in India in 1901, Russia in 1902, and Iran in 1906. In 1907, Edison's National Phonograph Company (never more than a bit player on the international scene) had subsidiaries in Europe, Australia, Argentina, and Mexico. By then, mass-produced musical records were available to consumers in Budapest and Sydney, Santiago and Beijing, Johannesburg and Jersey City. Although capitalization and manufacturing remained based primarily in the United States, Britain, France, and Germany, record-pressing plants opened in India in 1908 and China in 1914, and similar efforts were made with varying success in Australia in 1907 and Japan in 1911.[30]

Record labels soon succeeded around the globe, including the Lebanese Baidaphone label, for instance, which supplied customers across the Middle East, but had its records manufactured in Berlin from master disks produced in Beirut. By 1913, the Argentinean Discos Nacional label had its own studio and factory, and was selling millions of records a year in Argentina, while many of its tango recordings were also being issued in Europe under other labels.[31] The result was as much a matter of negotiating and circulating cultural difference as it was of homogenizing cultures or consumption. The popular success of recording helped to foster "a vast range of new urban popular musics" (A. Jones 2001, 54), adaptive indigenous expressions that flourished amid cultural politics at once local *and* global. By some accounts, the American Columbia label issued more "foreign" titles within the United States than it did other ones, so successful were its efforts to supply the nation's immigrant audiences and niche markets between 1908 and 1923 (Gronow 1982, 5).[32] Meanwhile, the Gramophone Company in India issued catalogs in Punjabi, Urdu, Hindi, Bengali, Tamil, Telegu, and Malayalam, at the same time that it employed at least one popular artist who recorded in English, Arabic, Kutchi, Turkish, Sanskrit, and Pushtu.[33]

What these examples suggest about media is far more interesting and complicated than the homogenization or Americanization of cultures, or the unparalleled purchase of the globalizing postmodern. Media help to *"organize and reorganize popular perceptions of difference within a global economic order,"* so that increasingly "one's place is not so much a matter of authentic location or rootedness but one's relationship to economic, political, technological, and cultural flows" (Curtin, 2001, 338). Increasingly, in other words, global media help to create a world in which people are not local only because of where they are or are from but also because of their relationships to media representations of localism and its fate. Even before the First World War, the experience of playing records and consuming the varied conventions of recording—including the varied patterns of commodification—turned the new medium of recorded sound into "something like the first global vernacular" (Hansen 1999, 68).[34] Here, I am drawing on Michael Curtin's description of television today and Miriam Hansen's account of Hollywood films in the "classical" period, but their points do hold nicely for early recorded sound and first-wave globalization.

Recorded sound remained new in the first years of the twentieth century in something of the same sense that digital communications remain new at the beginning of the twenty-first: widely perceived as technologically advanced and advancing, globally connected amid intense competition, unstinting hype, and increasingly open and extensive markets. Of course, there are differences between globalization now and globalization then as well as between different constructions of the new. The comparative study of media must be exactingly contrastive. Yet there are obvious parallels to be drawn too, and I think—it may be clear by now—that the early history of recorded sound holds a particular resonance for envisioning what can today be called the early history of digital media. Part of this resonance is superficial, but part of it involves the idea of history itself—what it means to experience a sense of history or historical fact, what it means to write the early history of anything, and what the histories of media specifically involve. In part because recorded sound developed in ways that its earliest promoters and audiences did not expect, and because digital networks have likewise developed in unanticipated ways, both cases offer a chance to cut across the technological determinism of popular accounts while at the same time allowing a more nuanced sense of how the material features of media and the social circulation of material things help variously to shape both meaning and communication. Media histories that lack this conjoined interest in the material and the historiographical have tended to dismiss or diminish the importance of phonographs in favor of electronic contemporaries, particularly telegraph and telephone networks, which so intuitively began to "dematerialize" communication along the trajectory that distributed digital networking today extends.[35]

At the broadest level, the initial development of recorded sound for improved business communications and its eventual incarnation as (at least primarily) a domestic amusement do suggest a number of immediate parallels to digital media. Like the transition from mainframe computers to PCs, the new medium became less centralized and expensive to use as well as more "personal" with better storage capacity. Like the text-based World Wide Web developed at the European Organization for Nuclear Research (CERN) and then transformed by the success of a more image-inclusive browser, Mosaic, written by programmers at the University of Illinois, the new medium of recorded sound was stripped of its research and development (R & D) past and became broadly commercialized. And like MP3 files and file-sharing technology for downloading music, the new medium distributed music in a new format, challenging existing market structures and provoking the bitter disputes over intellectual property that I have analyzed elsewhere.

Though suggestive, comparisons like these can also be pretty glib, and I want to dwell instead on another kind of parallel between recorded sound and digital media. This is a book less about sound than about text, less about the political economy of music than about the social experience of meaning as a material fact. Edison's phonograph inscribed in a new way, one that many of its first users evidently found mysterious. The inscriptions that Edison's phonograph made were tangible, portable, and immutable: records. But unlike more familiar inscriptions, they were also illegible. No person could read recordings the way a person reads handwritten scrawls, printed pages, or musical notes, or even the way a person examines a photograph or drawing to glean its meaning. Only machines could "read" (that is, "play") those delicately incised grooves. To top it off, Edison's phonograph seemed to inscribe or "capture" sound indiscriminately, capaciously—anything from noise to music—without regard for the speaker or the source. And it seemed to inscribe directly, without using ears, eyes, hands, a pencil, or an alphabet. The accounts rendered here of 1878 and 1889–1893 (chapter 1) and 1895–1910 (chapter 2) are in part a cultural history of the ways these new inscriptions were apprehended and commodified—that is, the ways these new inscriptions became gradually less mysterious as inscriptions and more transparent as forms of or aids to cultural memory, part of and party to the data of culture.

Digital media inscribe too, and they do so in what are mysterious new ways. (Mysterious to me, at least, and anyone else without an engineering background.) I see words written on my computer screen, for instance, and I know its operating system and other programs have been written by programmers, but the only related inscriptions of which I can be fully confident are the ones that come rolling out of the attached printer, and possibly the ones that I am told were literally printed onto chips that have been installed somewhere inside. At least inscriptions like printer output and microprocessor circuits share the properties of tangibility, portability, and immutability. The others? Who knows? I execute commands to save my data files—texts, graphics, sounds—but in saving them, I have no absolute sense of digital savability as a quality that is familiarly material. I have tended to chalk this up to the difference between the virtual and the real, without stopping to ponder what virtual inscriptions (N. Katherine Hayles [1999, 30–31] calls them "flickering signifiers") could possibly be.[36] Like the mysteries surrounding the inscription of recorded sound onto surfaces of tinfoil and then wax at the end of the nineteenth century, the mysteries surrounding the virtual inscription of digital documents are part of the ongoing definition of these new media in and as they relate to history. History "is written," Steve Jones (1999, 23) imagines, for instance, "in the electrons, generally, or [the] magnetic particles or pits and valleys that make up" different storage media. Like so many casual appeals to itty-bitty ones and zeros, there is an element of practical fantasy or useful fiction here that makes a difference to the emergent meanings of digital media.

Different inscriptions do make a difference. The sociologist Bruno Latour (1990) has demonstrated just how powerful inscriptions (his "immutable mobiles") are in the work of science. Scientists collect and circulate inscriptions, using some inscriptions—like electron micrographs, data sheets, lab notes, and cited articles—to produce others—such as grant applications and scientific papers for refereed journals. Other disciplines or types of inquiry work this way too. Classicists, for instance, work partly with inscribed archaeological artifacts (stone tablets, coins, and so on) and inscribed archival ones (papyrus, vellum, and paper; manuscripts, print editions, concordances, and monographs). And of course, society at large depends on oodles of different inscriptions, everything from street signs, newspapers, and videos, to medical charts, price tags, and paperbacks. The relative functions or merits of different sorts of inscriptions can be difficult to parse, particularly if one is unfamiliar with the social contexts in which they circulate. There are inscriptions that make sense in broad contexts (any adult knows how a ten-dollar bill works, for example) and others that make sense only in exactingly narrow contexts (like a baby picture, a dry-cleaning ticket, or the tiny accession numbers painted by a museum curator onto a rare specimen). Whole new modes of inscription—such as capturing sounds by phonograph in 1878, or creating and saving digital files today—make sense as a result of social processes that define their efficacy as simultaneously material and semiotic. A computer engineer can explain how digital files really are created and saved, but I would insist that the vernacular experience of this creatability and savability makes at least as much difference to the ongoing social definition (that is, the uses) of new, digital media.

Because they are at some level material, one important quality that all inscriptions share is a relationship with the past. Whether scribbled down just a second ago or chiseled into stone during the sixth millennium BCE, whether captured in the blink of a shutter or accumulated over months and years of bookkeeping, inscriptions attest to the moments of their own inscription in the past. In this sense, they instantiate the history that produced them, and thus help to direct any retrospective sense of what history in general is.[37] For example, the history of the Salem witch trials is known largely because people at the time wrote about them. These documents contain legible information, but they also carry plenty of other data by virtue of their materiality—their material existence and material or forensic properties. Historians today read the Salem documents, of course, yet they also "read the background"; they analyze the written words, but they also assess the look, feel, and smell of the paper, sometimes without even realizing they're doing so.[38] A shared sense of writing, of what can be written down and what cannot, also helps make them comprehensible in a lot of subtle ways. A whole social context for and of writing existed then in Massachusetts, and a related context presently exists, although today's tacit knowledge of writing includes influential details about what writing isn't: it isn't like photography; it isn't like sound recording. Modes of inscription that Salem witches and divines could never have imagined in the seventeenth century are now subtly and unavoidably part of the way that seventeenth-century inscriptions are understood.

This means that media are reflexive historical subjects. Inscriptive media in particular are so bound up in the operations of history that historicizing *them* is devilishly difficult.

There's no getting all of the way "outside" them to perform the work of historical description or analysis.[39] Our sense of history—of facticity in relation to the past—is inextricable from our experience of inscription, of writing, print, photography, sound recording, cinema, and now (one must wonder) digital media that save text, image, and sound. The chapters that follow are in one sense argumentative examples of exactly this. They demonstrate how new modes of inscription are complicated within the meaning and practice of history, the subjects, items, instruments, and workings of public memory. Inquiring into the history of a medium that helped to construct that inquiring itself is sort of like attempting to stand in the same river twice: impossible, but it is important to try, at least so the (historicity of the) grounds of inquiry become clear.

How does the same sort of reflexivity complicate today's new media? How is doing a history of the World Wide Web, for instance, already structured by the Web itself? How is digital inscription, with its mysteriously virtual pages and files, part of an emergent, new sense of history for the digital age? Chapters 3 and 4 pursue questions like these in different yet complementary ways. Chapter 3 looks at some of the earliest instances of digitally networked text. It asks how creators and users of the ARPANET, the precursor to the Internet, experienced computer networks as requiring or related to inscription. What was the larger economy of inscription and inscriptiveness within which they experienced digitally networked text? What were the documents amid and against which digital ones might have been defined? Like chapter 1 in its focus on 1878 and 1889–1893, chapter 3 opens a narrow window, 1968–1972, in order to glimpse a new medium at its newest. Then, like chapter 2, chapter 4 broadens this prospect by focusing on later, more popular uses of still-emergent digital media. It asks how history is represented on the World Wide Web and how the Web is being used to represent its own history. Further, it asks how using the Web may be prompting users to underlying assumptions about the new and the old, about a sense of time, a sense of present and past, and even a sense of ending. My idea is that this last question, about using the Web, is the one that reveals just how linked the first two are: history *on* the Web and history *of* the Web. These are not identical, of course, but they are inextricable.[40]

Like the missionaries who wrote histories of the Americas seemingly moments after stepping off their ships from Europe in the sixteenth and seventeenth centuries, a good number of people have already written histories of the Internet and the World Wide Web. Although the first Web server only went online in 1990, for instance, "The orthodox accounts ([Vannevar] Bush to [Doug] Engelbart to [Ted] Nelson to everything else)," admits Michael Joyce (2001, 211), have already taken "on the old testamentary feel of the Book of Numbers: 'Of the children of Manasseh by their generations, after their families, by the house of their fathers.'"[41] The Moses or Edison of these patrilineal accounts tends to be Timothy Berners-Lee, the computer scientist at CERN who wrote and released the Web's basic architecture, prompted the first generation of browsers, and now heads the World Wide Web Consortium (W3C) based at MIT.[42] He and his colleague, Robert Cailliau, pitched the Web to their employer as an information management tool for CERN's own continued work in particle physics. Chapter 4 will look further into how this history of the Web is being told, as well as how the Web appears in some respects to resist history.

Beyond CERN, the broader physics community made early use of the World Wide Web. For instance, the library at the Stanford Linear Accelerator Center (SLAC) soon offered Web-based access to "preprints"—articles that are on their way through the peer-review process, but that haven't appeared in print or electronically yet with the final imprimatur of a refereed journal. The new accessibility of preprints made them not more authoritative but certainly more integral to the work of physicists. The practice of doing physics (like doing classics, as it happens) changed in keeping with the accessibility and abundance of what had before been inscriptions that circulated slowly and in narrow contexts.[43] Elsewhere on the disciplinary map, doing art history has also changed in similar ways, but it changed first in the early twentieth century with the advent of slide lectures as a defining pedagogical practice. As Robert Nelson (2000, 417, 422) explains, the slide in an art history lecture gets referred to and treated not as a "copy of an original, but as the object itself," so that "arguments based upon slides alone are persuasive, even if the evidence only exists within the rhetorical/technological parameters of the lecture itself" (as, for instance, "when objects of greatly different sizes and from unrelated cultures are regarded as comparable because they appear side by side in the slide lecture"). According to Nelson, the result was a gradually more inductive and positivistic discipline; because or as part of the widespread adoption of slides in lecturing, artworks became self-evident facts in a new way.

There is an anachronistic or before-the-fact echo of Hayles's flickering signifier here in the lecture hall, with new layers of semiotic process between art students and their subjects. But what these thumbnail histories of disciplines help to suggest more broadly is that the properties, accessibility, and abundance of inscriptions matter to their facticity, not what's true or false but rather what counts as knowledge and what doesn't, what questions seem interesting and important to ask.[44] And if the facticity and practices of doing physics and doing art history have changed in accordance with changing modes of inscription, it seems reasonable to think that the disciplinary practice of doing media history is changing with the media that it does history to.

I *The Case of Phonographs*

1 New Media Publics

1878: Tinfoil

Like any new medium, recorded sound could not but emerge according to the practices of older media. Edison stumbled on the idea of sound recording while working on telephones and telegraphs during the summer and fall of 1877, and communication devices like these provided an initial context for defining the phonograph. To the inventor and his contemporaries, the phonograph meant what it did because of the ways it might resemble and—particularly—because of the ways it might be distinguished from existing machines. As Edison (1878, 527) put it, what made the phonograph so different was its "gathering up and retaining of sounds hitherto fugitive, and their [later] reproduction at will." What had always been lost, what had previously fled, could now be gathered up or "captured" and stored for future use. And of course, the fugitive sounds captured by the phonograph meant what they did because of the ways they might resemble and—particularly—because of the ways they had to be distinguished from the only other snare available: inscriptions made on paper.[1]

The torrent of accounts in the press that ensued all suggest that the first phonographs were initially understood according to the practices of writing and reading, particularly in their relation to speaking, and not, for instance, according to the practices or commodification of musical notation, composition, and performance. The really remarkable aspect of the "speaking phonograph" or "talking machine" arose, as some of its earliest observers marveled, in "literally making it read itself." A record was made and then played, as if, "instead of perusing a book ourselves, we drop it into a machine, set the latter in motion, and behold! The voice of the author is heard repeating his own composition."[2] Hidden here are what James Lastra (2000, 6–7) has identified as the two "tropes for understanding and normalizing" new media: one of inscription, and the other of personification.[3]

Assisting in the public apprehension of the phonograph as a textual device and a metaphoric author and reader were structural as well as functional comparisons: those first recordings were indented on sheets of tinfoil. Like some celebrated author-orator, Edison's phonographs went on the lyceum circuit during the summer of 1878, publicly "writing on" and "reading" or "speaking from" their sheets.

Audiences at the phonograph demonstrations were both enthusiastic and skeptical, responding simultaneously to the unprecedented marvel of recorded sound and the unreasoned hype surrounding its early, imperfect demonstration. Those early recordings were faint, flimsy, and full of scratchy surface noise. After the initial fanfare died down, the public had little opportunity to experience recorded sound again firsthand until 1889–1893, when improved phonographs playing wax cylinder records were exhibited and started to become available for demonstration and as nickel-in-the-slot amusements, playing prerecorded musical records in public, urban locales and as diversions at fairs and summer resorts. These machines were met with more enthusiasm (lots of nickels) and skepticism, since neither the mechanisms nor the wax recordings delivered much of what they promised. Nonetheless, musical entertainment had become a presumed function of the new medium, though Edison and others continued to promote the devices as business machines for taking dictation.[4]

This chapter is about these early incarnations of the new medium. It is about the public life of phonographs at a time when publics and public life were the incumbent structures of print media. Americans of the day thought of themselves as constituents of a nation and nationally constituent localities according in part to their ritualized collocation as readers of a shared press, as private subjects within the same vast, public, and calendrical circulatory system for printed matter.[5] Even the most disadvantaged could occasionally self-enroll as members of society by dint of literacy acquisition. Frederick Douglass ([1892] 1976, 89) called the antiabolitionist Baltimore newspapers "our papers," when he recalled reading them as a youth, verbally including himself as a constituent of the very public, the very "us," that had attempted systematically to deny his humanity. Because it curiously pertained to "papers," the early history of recorded sound had something to do with "us" and "our." One of my points is that all new media emerge into and help to reconstruct publics and public life, and that this in turn has broad implications for the operation of public memory, its mode and substance. The history of emergent media, in other words, is partly the history of history, of what (and who) gets preserved—written down, printed up, recorded, filmed, taped, or scanned—and why.

The interrelated meanings of print and public speech have been the subject of intensive study by scholars of early U.S. literary history.[6] Such accounts, however, seldom encompass the nineteenth or twentieth centuries, when the liberal nation-state was more firmly in place.[7] As James Secord (2000, 523) explains in his study of early Victorian print production and reception, the West experienced an "industrial revolution in communication" during the mid-nineteenth century, stemming in large part from interconnected changes to the technology of print production and to "the form of public debate." At the broadest level, notes Secord, "the power of print lies in the [shared] assumption of its fixity," and in the nineteenth-century United States as in Britain, print came unglued.[8] The relative instability of nineteenth-century print may be glimpsed first in the sheer volume of print media, in what one observer remarked as "books in shoals [and] journals by the score" (Farmer, 1889, vii), but particularly in the profusion of cheaper monthly, weekly, and daily periodicals. By one account, there were 8,129 local newspapers in the United States and its territories in 1876.[9] Yet quantity is far less suggestive than quality. Under the U.S. "exchange" system (the subsidized postal swapping of issues across the country), newspapers and periodicals reprinted voraciously. Local papers culled each other's pages, assembling a national press. Meanwhile, without international copyright strictures, U.S. publishers pirated fervidly, particularly from the British press. The result was an unfixity of print perhaps unprecedented since the seventeenth century.[10]

Part of the most basic connection between print and *fact*—that transparent Enlightenment logic that operates (what Michel Foucault identified as) the "author function," and that lionizes textual authenticity and legitimates textual evidence[11]—eroded in practice: readers know today how frustrating it is to pick up an edition, even an authorized "complete works" from the period, or a newspaper column, not to say a copy of a British novel published piratically in the United States, and receive little or no indication of the provenance of the work it presents. Where and when was it originally published, and by whom? Whether newspaper "exchange editors" snipped it off or larcenous publishers obscured it as a matter of course, provenance came loose from texts amid the fertile chaos of industrial print production. Sometimes the most elaborately authenticated texts—framed by testimonials and prefaced by respected authorities—were encountered as the least trustworthy.[12] Even the Bible seemed newly unstable, widely and differently printed in English amid accelerated efforts toward revision and retranslation.[13] And if it was impossible to tell where a text had originally come from, it was increasingly difficult to tell where a text was eventually going. As literacy and reading publics expanded, potential readings did too. To make this point narrowly, studying literary texts before this period had much more to do with appreciation than with interpretation, because "a gentleman could be depended upon to understand intuitively what" such a text meant (Graff and Warner 1989, 4).[14] Once mass literacy met cheap editions, things were different. With

provenance loose (bibliographically), reception unpredictable (sociologically), and questions of authorial ownership vexed, the fixity of print was in jeopardy and the social meanings of print were in flux.

No less fluid were the meanings of public speech in the nineteenth-century United States, though these meanings are notoriously hard to get at. The work of "governing the tongue" or speaking properly had long before ceased to be a matter of legality or the evidence of salvation, and had become a more subtle, more personalized "code distinguishing the refined from the vulgar," an "exercise in self-control" (Kamensky, 1997, 190, 182). By Reconstruction, keeping a "civil" tongue had nothing at all to do with the civil authorities—that is, with the notable exceptions of schoolchildren (particularly Native American and nonnative ones, whose "uncivilized" speech tended to be circumscribed) and President Andrew Johnson, who was impeached in 1868 partly because of allegedly "intemperate, inflammatory, and scandalous harangues" as well as "loud threats and bitter menaces" uttered in public against Congress. For others who were neither president nor pupil, speech possessed a less specific, if still public, valence.[15]

In bourgeois circles, "calls for plain speech remained a key component of demands for a democratic culture" (Lears 1994, 53), as successive generations of verbal critics fulminated against a decline in the standards of American usage.[16] The verbal critics were caught in a telling dilemma: they pointed to U.S. newspapers as their evidence of declining standards in usage, but they also pointed to them as the single most important cause of decline. So-called newspaper English was the "arch villain" in their campaign for rectitude (Cmiel 1990, 134–135). The logical conundrum of blaming as cause that which is also ascribed as effect was overlooked, I suspect, because readers experienced the newspaper as contradictorily bivalent, as printed speech. The fleeting currency of news, the ephemerality of the papers, rendered them more like speech acts and less like print artifacts, while their tangibility conversely rendered them "hard" evidence in black and white. Materially, newspapers were print. Legally, however, they tended to resemble vocal performances more than they did authored forms. According to a precedent established by the U.S. Supreme Court in 1829, nothing with "so fluctuating and *fugitive* a form" could possess copyright, which the Constitution reserved for "more fixed, permanent, and durable" expressions.[17] In this context, Edison's "capture" of "fugitive" sounds onto sheets of tinfoil offered nothing less than the renovation of newsprint.

Because the phonograph demonstrations of 1878 and the nickel-in-the-slot business of the early 1890s variously interrogated the normal habits of writing, reading, and speaking, they offer an opportunity to gauge the meanings of print media during the late nineteenth century. For sure, these same meanings may also be glimpsed in other, attendant experiences of text—for instance, in the ongoing legal construction of authorship, in the changing political economies of publishing, the additional subjectivities of late-century literacy, the shifting character and institutional status of criticism, and so on. The list is long. Against it, I will suggest here that tinfoil records, in particular, can profitably be read as *foils* in the literary or schoolbook sense. They were historical characters important in their pairing function, defining by contrast, relative and mutually opposed to other characters—characters like authors, readers, publishers, and critics, but even more particularly the characters who "spoke" between quotation marks from the pages of U.S. newspapers. Subsequent adaptations of the phonograph into a musical amusement device involved corresponding adaptations to what Benedict Anderson (1991) has called "print capitalism," the cultural economies of print circulation and consumption that so powerfully helped to articulate a U.S. public, an "us," with "our" "own" national tradition and aspirations.

Along with the surrounding publicity, tinfoil records offered a profound and self-conscious experience of what "speaking" on paper might mean. Judging at least from the coincident popularity of verbal criticism, or the coincident quarrels between philologists and rhetoricians over the appropriate study of language, Edison's invention was less a causal agent of change than it was fully symptomatic of its time. Related issues of speaking on paper were also raised in contemporary promotions of simplified spelling, rampant competition between different shorthand systems, and published dialect and regional literatures as well as scholarly worries about the "correct" pronunciation of Latin, the probable pronunciation of English by Chaucer and Shakespeare, and the appropriate "collection" of non-Western tongues, "authentic" black spirituals, folktales, and English ballads.[18] This chapter tells the story of both a few, fragile sheets of tinfoil and then a short-lived nickel-in-the-slot amusement because these stories offer another, modest opportunity to look into the concerns of their time. In particular, the early history of sound recording makes visible the ways in which new media emerge as local anomalies that are also deeply embedded within the ongoing discursive formations of their day, within the what, who, how, and why of public memory, public knowledge, and public life.

In January 1878, Edison signed contracts assigning the rights to exhibit the phonograph (and to make clocks and dolls), reserving for himself the right to exploit its primary dictation function at a later time. Exhibition rights went to a small group of investors, some of them journalists and most of them involved already in the financial progress of Alexander Graham Bell's telephone. In April, they formed the Edison Speaking Phonograph Company,

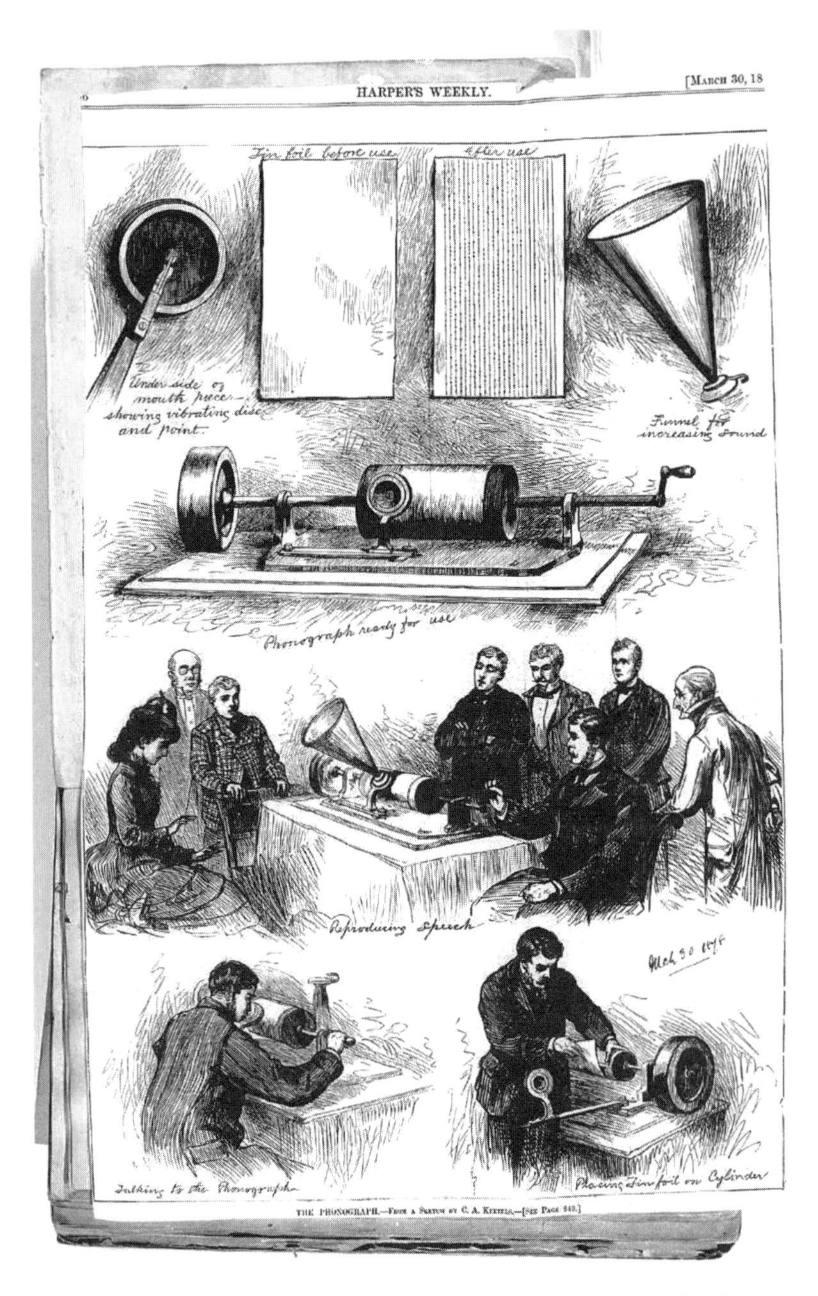

Figure 1.1 Tinfoil phonograph from *Harper's Weekly,* March 30, 1878. Examples of tinfoil, used and unused, appear at the top of the page. (*Source:* Thomas A. Edison Papers Digital Edition.)

and they hired James Redpath as their general manager in May. A former abolitionist, Redpath had already helped to transform the localized adult-education lecture series of the early American lyceums into more formal, national "circuits" administered by centralized speakers' bureaus. He had just sold his Redpath Lyceum Bureau, and came to the phonograph company with a name for exploiting "merit" rather than what his biographer later dismissed as "mere newspaper reputation" (Horner 1926, 227, 185). The distinction was a blurry one throughout the ensuing months of phonograph exhibitions. Like so much entertainment then and now, phonograph exhibitors capitalized on novelty. Novelty wore off, of course, though it would be about a year until they had "milked the Exhibition cow pretty dry," as one of the company directors put it privately in a letter to Edison.[19]

The Edison Speaking Phonograph Company functioned by granting regional demonstration rights to exhibitors. Individuals purchased the right to exhibit a phonograph within a protected territory. They were trained to use the machine, which required a certain knack, and agreed to pay the company twenty-five percent of their gross receipts. This was less of a lecture circuit, then, than it was a bureaucratic one. Phonograph exhibitors were local; what tied them together into a national enterprise were corporate coordination and a lot of petty accountancy. Paper circulated around the country—correspondence, bank drafts, and letters of receipt—but the people and their machines remained more local in their peregrinations, covered in the local press, supported or not by local audiences and institutions in their contractually specified state or area. Admission was set at twenty-five cents, though soon there were exhibitors cutting that to a dime. Ironically, no *phonographically* recorded version of a phonograph exhibition survives. Despite wishful reports to the contrary, the tinfoil records did not last long and were difficult to replay. Instead, the character of these demonstrations can be pieced together from the many accounts published in local newspapers, letters mailed to Redpath and the company, and a variety of other sources, which include a minstrel burlesque of the exhibitions titled *Prof. Black's Phunnygraph, or Talking Machine.*

While the Edison Speaking Phonograph Company was getting on its feet, Edison along with his friends and associates made some of their own exhibitions, setting a pattern for Redpath's exhibitors to follow and raising the expectation of company insiders. Edison himself appeared *gratis* before an audience at the National Academy of Sciences in Washington, DC. The exhibition included an explanation of how the machine worked, and then Edison's assistant "sung and shouted and whistled and crowed" into its mouthpiece. As the *Washington Star* reported the next day, "There was something weird and uncanny about the little machine," as it then "expressed itself," and "the same sounds floated out upon the air faint but distinct." Later on, the demonstration was marred for an instant when the tinfoil ripped, but excepting this interruption, "the awful hush that preceded the phonograph's remarks was broken [only] by its far off utterances." As he had during the preceding months, Edison gladly granted an interview to reporters. He vaunted the promise of recorded sound for preserving speech or recording a famous diva, and reported that the American Philological Society had requested a phonograph "to preserve the accents of the Onondagas and Tuscaroras, who are dying out." According to the *Star,* Edison said, "One old man speaks the language fluently and correctly, and he is afraid that he will die. You see, one man goes among the Indians and represents the pronunciation of their words by English syllables. Another represents the same words differently. There is nothing definite. The phonograph will preserve the exact pronunciation."[20] Edison implied a distinction between one man and another, between "men" and "the Indians," which he mapped against a distinction between the contrivances of representation and the natural fluidity of spoken language.

Edison's friend Edward H. Johnson gave the first demonstrations for hire, pairing the phonograph with versions of the telephone. He charged theater managers one hundred dollars a night during a tour of New York State in January and February. Not all of his performances recouped the hundred dollars, but in Elmira and Cortland "it was a decided success," he claimed, and the climax of the evening was "always reached when the Phonograph first speaks." "Everybody talks Phonograph," Johnson reported, on "the day after the concert and all agree that a 2nd concert would be more successful than the first." Johnson's plan was a simple one. He categorized the fare as "Recitations, Conversational remarks, Songs (with words), Cornet Solos, Animal Mimicry, Laughter, Coughing, etc., etc.," which would be "delivered into the mouth of the machine, and subsequently reproduced." Johnson described getting a lot of laughs by trying to sing himself, but he also attempted to entice volunteers from the audience or otherwise to take advantage of local talent.[21] A month later, Professor J. W. S. Arnold half filled Chickering Hall in New York City, where his phonograph "told the story of Mary's little lamb," and then like Johnson's phonograph, rendered a medley of speaking, shouting, and singing. At the end of the evening, Arnold distributed strips of used tinfoil as souvenirs, and there was reportedly "a wild scramble for these keepsakes."[22]

By the time the Speaking Phonograph Company swung into action, New York City, at least, had been pretty thoroughly introduced to Edison's invention. Redpath complained (perhaps facetiously) that the city had endured more than three hundred demonstrations by the time of his own short season at Irving Hall, which was tepidly received. The company's contracted exhibitors sought less jaded audiences. Across the Hudson in New Jersey, for instance, a journalist named Frank Lundy owned demonstration rights, and his activities can be gauged in the local press. For example, Lundy came through Jersey City, New Jersey, in mid-June, where he gave one exhibition at a Methodist Episcopal church ("admission 25 cents"), and another at Library Hall as part of a concert given by "the ladies of Christ Church." Both programs featured musical performances by community groups as well as explanations and demonstrations of the phonograph. Lundy reportedly "recited to" the machine, various "selections from Shakespeare and Mother Goose's melodies, laughed and sung, and registered the notes of [a] cornet, all of which were faithfully reproduced, to the great delight of the audience, who received pieces of the tin-foil as mementos." Poor Lundy's show on June 20 had been upstaged the previous day by a meeting of the Jersey City "Aesthetic Society," which met to wish one of its members bon voyage. Members of the "best families in Jersey City" as well as "many of the stars of New York literary society" were reportedly received at Mrs. Smith's residence on the eve of her departure for the Continent. One New York journalist had brought along a phonograph, and occupied part of the evening recording and reproducing laughter as well as song, along with a farewell message to Smith, and a certain Miss Groesbeck's "inimitable representation" of a baby crying. Of these recorded cries, "the effect was very amusing," and the journalist "preserved the strip" of foil, saving the material impressions of what were reported as Groesbeck's "mouth" impressions.[23]

Further from Edison and the company's orbit, demonstrations of the phonograph appear to have been similar affairs. In Iowa, George H. Iott of Des Moines owned exhibition rights. The *Iowa State Register* reported "The Phonograph in Des Moines" on July 3, 1878. Iott's demonstrations ran morning, noon, and night in one of the city's commercial blocks, for any who wished to visit "the wondrous machine of iron, steel and foil that can be made to talk, whistle, sing, crow, laugh or make any other vocal sound." Like Groesbeck in New Jersey, Iowans got a chance to make their own records. As the *Register* put it, "Quite a number of our people sung and talked to this phonograph yesterday." A lawyer recorded an argument to the court, another man recorded the Lord's Prayer, "several ladies sang to it," a professor spoke to it in foreign languages, and the machine "repeated" them all. The foreign languages sounded funny, and Iott's own renditions of "John Brown's Body" and "Whoop Her Up, Eliza Jane" proved to be crowd-pleasers. Two weeks later, during a horrible heat wave, the phonograph appeared in Dubuque. Iott probably operated it there too, though the local *Daily Times* called it only "Edison's" phonograph. In advance of the Dubuque exhibitions, someone else had already offered "a specimen of tinfoil," for display in town, it "being a section of the band which encircled the cylinder of an Edison phonograph," presumably at a public demonstration someplace else. Even without the phonograph to play it on, "the solidified, or rather materialized effects of voice" visible on the foil were "indeed a curiosity."[24]

Though there was obviously some variety among them, phonograph exhibitions all shared a similar form and content. Their structure was self-fulfilling, interactive, and based on a familiar rhetoric of educational merit. Lecturers introduced Edison's machine as an important scientific discovery by giving an explanation of how it worked, and then the "how" was confirmed in successive demonstrations of recording and playback. Audiences were edified, and they were entertained.[25] The demonstrations first involved recording a hodgepodge of sounds that ranged from shouting, singing, and recitation, to noises like coughing, laughing, and crowing, all by the lecturer and all reproduced by the machine to the silent assembly. Then, a few audience members got a chance to record sounds of their own, which were also reproduced. By making recordings themselves, audiences became part of the demonstration and party to the new medium—instrumental to the claims of each.

The exhibitions formed an intricate and spontaneous response to the contemporary cultural order of which they were also functioning ingredients. After the contested presidential election of 1876 and the great railroad strike of 1877, with the economy still recovering from a deep depression, the phonograph helped to encourage a "renewed optimism about America's future," serving as an early palliative within what historian Robert Wiebe (1967) has called the post-Reconstruction "search for order."[26] Though their meanings for different audiences must have been to some degree different, the exhibitions fostered consensus. They were local experiences—many audience members probably knew each other—yet they had extralocal significance. Like the exchange columns and wire stories of the local press, phonograph exhibitions pointed outward, toward an impersonal public sphere comprised of similarly private subjects. Audiences in the meanest church basements were recorded just like audiences in the grand concert halls of New York, Chicago, and New Orleans. In their very recordability, people were connected. Audience members might imagine themselves as part of an up-to-date, recordable community, an "us" (as opposed to some imagined and impoverished "them"), formed with similarly up-to-date recordable people they didn't know.

Phonograph exhibitions drew audiences together by flattering them in two different ways. On the one hand, the exhibitions offered tacit participation in technological progress to everyone in attendance. Audiences could be up to the minute, apprised of the latest scientific discovery, in on the success of the inventor whom the newspapers were calling the "Wizard of Menlo Park" and the "Modern Magician." Together, audiences might imagine and salute all of the useful functions the phonograph would serve, once Edison had improved the device as promised. On the other hand, the exhibitions provided a playful and collective engagement with good taste. In making their selections for recording and playback, exhibitors made incongruous associations between well-known lines from both Shakespeare and Mother Goose, between talented musicians and hacks like Edward Johnson, between animal and baby noises and the articulate sounds of speech. Audiences could draw and maintain their own distinctions, laugh at the appropriate moments, recognize impressions, and be in on the joke. They could participate together in the enactment of cultural hierarchy.

Cultural hierarchy was enacted partly through carnivalesque gestures—body sounds or animal noises—the negative of bourgeois identity, newly contained, captured, by the mimetic device. (One of the first Edison kinetoscope films offered, similarly, *The Record of a Sneeze,* in 1894.) Edison had proposed that his machine might preserve "our Washingtons, our Lincolns, [and] our Gladstones"—what Matthew Arnold called "the best that has been thought and *[literally]* said"—and yet the public capaciousness of the phonograph seemed to ask for the low, the other, and the infantile—the pro- or protosemiotic—all performed cathartically within the respectability of the middle-class lecture space and its rational, technocratic weal. Thus, recording was from the outset a complex "studio art," in the words of Jonathan Sterne, in which "both copy *and* original are products of the process of reproducibility" (2003, 236, 241; emphasis added).[27] Confronted by a machine that preserved and repeated speech, individuals reduced their own expressions to rote. They repeated bits that were already often repeated: a prayer, a lyric, a snippet from a common vocabulary of quotations (from Shakespeare), or a piece from a common past (from the nursery). They mimicked to the machine they knew would mimic them mimicking. They made themselves fully its subjects, recording themselves by phonograph at the same time that they acknowledged its rote "memory" by comparing it to theirs.

Something of the same carnival circulated in the press, where the phonograph was hailed sincerely as the most wondrous scientific invention of the age, but was also the source and butt of jokes. The most frequent were misogynist gibes, maintaining the masculinity of publics and public speech by assigning private and aberrant speech to women, gossips, harpers, nags, talking machines that never require any tinfoil, and so on. Other jokes proposed new, related devices for inscription: the smellograph for recording and reproducing odors, the nip-ograph for recording and reproducing inebriation, and so forth. Like the great variety of sounds recorded during demonstrations, such jokes hint at the hugely varied contexts within which public speech acts made sense as bodily cultural productions. If phonographs were "speaking," their functional subjects remained importantly diffuse among available spoken forms: lectures and orations as well as "remarks," "sayings," recitations, declamations, mimicry, hawking, barking, and so on. The sheer heterogeneity of public speech acts should not be overlooked any more than the diversity of the differently public speakers whose words more and less articulated a U.S. public sphere. The nation that had been declared or voiced into being a century before remained a noisy place.

More specific audience response is difficult to judge. There were some parts of the country that simply were not interested. Mississippi and parts of the South were experiencing one of the worst yellow fever epidemics in recent memory. Audiences in New Orleans were reportedly disappointed that the machine had to be yelled into in order to reproduce well, and there were other quibbles with the technology once the newspapers had raised expectations to an unrealistic level. Out in rural Louisiana, one exhibitor found that his demonstrations fell flat unless the audience heard all recordings as they were made. Record quality was still so poor that knowing what had been recorded made playback much more intelligible. Redpath spent a good deal of energy consoling exhibitors who failed to make a return on their investments, but he also spent his time fielding questions from individuals who, after witnessing exhibitions, wrote to ask if they could secure exhibition rights themselves. To one exhibitor in Brattleboro, Vermont, Redpath wrote sympathetically that "other intelligent districts" had proved as poor a field as Brattleboro, but that great success was to be had in districts where "the population is not more than ordinarily intelligent." Some areas of the country remained untried, while others were pretty well saturated, like parts of Pennsylvania, Wisconsin, and Illinois.[28]

But one response in particular seems to have been commonplace: audience members took souvenir scraps of indented foil home with them when the exhibitions were over. These partial and primitive records were in some way meaningful to the women and men who sought them, and who were probably asked at the breakfast table the next morning, "What does it say?" Without the phonograph for playback, the tinfoil records of course said nothing. Yet at the same time, as a few lines in the morning newspaper might help to report, the very same records said *something*. These "materialized effects of voice" were "indeed a curiosity": they were talismans of print culture, pure "supplement," in the language of literary study today, illegible and yet somehow textual, public, and inscribed. Exhibitors wrote back to the company for more and more tinfoil. The company kept a "foil" account open on its books to enter these transactions. Pounds of tinfoil sheets entered into national circulation, arriving in the possession of exhibitors only to be publicly consumed: indented, divided, distributed, and collected into private hands, and then saved. This saving formed a totally new experience of savability as well as the preservative effects of tinfoil. That is, tobacco and cheese were sometimes sold in tinfoil, but the commercial availability of foil for wrapping and saving leftovers would come much later.

Confirming much about the phonograph exhibitions was a "colored burlesque on the phonograph" titled *Prof. Black's Phunnygraph, or Talking Machine.* Frank Hockenbery's (1886) skit offers a comment on the phonograph lectures that it lampoons. The term *burlesque* did not then denote striptease as much as it indicated a topical, risqué comedy, full of witticisms pointed at events of the day, and the butt of Hockenbery's burlesque are the phonograph exhibitions of 1878. As a "colored" burlesque, *Prof. Black's Phunnygraph* also taps fifty years of blackface minstrelsy in its makeup. This was the era of the so-called mammoth minstrel shows, touring troupes of forty to sixty performers, and Hockenbery's *Phunnygraph* was probably intended as an interlude in one of these racist pageants, since its concluding stage directions call for a minstrel staple, "moving to half circle, [and as] soon as half circle is struck, begin negro chorus or plantation melody. Minstrel business."[29]

Whatever its origins and performance history, *Prof. Black's Phunnygraph* takes aim at phonograph exhibitors like Johnson, Lundy, and Iott. The burlesque develops in the form of a lecture "on de Phunnygraph, or Talking Machine, as she am called by de unsophisticated populace." "Professor" Black is its "sole inwentor, patenter, manufacturer an' constructioner," although there is passing and disparaging mention of a "Billy Addison." Professor Black is a character adapted as much from medicine shows as from men like Edison or his associate Arnold. A connection to patent medicines is made abundantly clear by the scenery, which consists in part of a sign announcing the lecture within the play:

Admittance
Adults . . . . 10 cents
Children . . . . ½ dose

The rest of the scenery is "a dry goods box large enough to hold three persons" to which has been attached a "sausage-grinder on top with crank," a household funnel, and "slips of white paper to run into [the] grinder to talk on." In the circumstances of its demonstration, the phunnygraph closely resembles the phonograph it mocks. Professor Black's lecture follows the formula of so many Edison Speaking Phonograph Company exhibitions. The professor explains how the machine works and then gives a demonstration. Just like the real phonograph exhibitors, the professor tries to record talk, recitations, animal mimicry, whistling, and song. But his phunnygraph is really just three people hiding in a box, and it proves impossible for them to remember and accurately repeat the words and noises that the professor shouts into the kitchen utensils on top of the box.

Professor Black fails in his efforts to "talk on[to]" paper. His three accomplices (literally) draw a blank. The substitution of empty "white paper" for tinfoil in the opening stage directions again identifies the phonograph as a kind of writing instrument and mnemonic machine at the same time that it serves to underscore that souvenir tinfoil sheets remained vexingly illegible, blank to the eyes of readers. Aural culture could suggestively be saved for posterity, but saving threw the experienced norms of savability into question.

If the tinfoil souvenirs were unreadable, they were also unauthored in important ways. Repeated explanations told how air was set in motion by the human voice, moving a diaphragm and a stylus, which in turn caused indentations in the foil. The author's voice authored. The *Iowa State Register* called it "incomprehensible." Added to what may have been a vague sense of how the machine worked, the phonograph exhibitions did plenty to confuse authorial agency in other ways. In each, the exhibitor was the author in evidence, responsible for manipulating the machine along with its stylus and sheets, and principally in charge of the substance of the recordings made. Audience members shared the exhibitor's authorial role when they made their own recordings. The machine, so readily personified, was an authorial subject too, to the extent that later exhibitors were chided to remember, "The audience is always more anxious to hear the machine than to hear you."[30] Finally, Edison remained author, a paradigmatic self-made man and maker, everywhere associated with his invention, sometimes occluding the phonograph's local exhibitor in the press. He received a royalty of 20 percent from the Edison Speaking Phonograph Company's net receipts, and the company tried to get Edison to control his image. One of the company's directors figured that they could sell a hundred thousand souvenir photographs of the inventor, if only he would carefully limit his exposure to photographers.[31] This gambit probably never came off, but the initial estimate of a hundred thousand was plausible; four years later, it was easy to sell eighty-five thousand souvenir photographs of Oscar Wilde as he made a lecture tour around the United States.[32]

The analogy so abundantly apparent between souvenir photos and souvenir tinfoil remained curiously absent from most early accounts of phonograph exhibitions. Though a few individuals had wondered before about the possibility of an "acoustic daguerreotype," and many were soon willing to compare Edison's phonograph to "its sister instrument, the camera," their analogy simply does not surface in accounts of 1878 with any great regularity.[33] If anything, Hockenbery's facetious analogy to patent medicines had more weight. Edison and others noted the miraculous power of his invention to "bottle up" speech for posterity, and bottling adapted well to the partially carnivalesque tenor of the exhibitions. Indeed, medicine shows also offered multiauthor fun and focused on inscrutable matter. As an itinerant "professor" lectured on the benefits of his pills or elixirs, the showman's company entertained the audience; James Whitcomb Riley remembered his youthful days sketching busts of Shakespeare and writing bad puns on two chalkboards while a "doctor" C. M. Townsend droned on about his miraculous Wizard Oil in 1875.[34] Like so many bottles of elixir, tinfoil promised much but delivered little. What the phonograph exhibitors were selling besides novelty was amazing curative potential as yet unrealized. The speculative future functions of Edison's device suggested social and cultural ills, or lacks that might soon be as fugitive as the fugitive sounds the phonograph would capture, once and for all.

The tinfoil was thus the site of enormous tension. Phonograph exhibitors ran through pounds of it, and audiences scrambled for keepsakes. In their sonic "capture" and later mute evocation of public experience, pieces of foil in private hands must have formed souvenirs of curious power. Like all souvenirs, they were belongings that vouched for belonging. They were artifacts that vouched for facts. Publicly made and privately held, their very material existence offered a demonstrable continuity of private and public memories, hard evidence of shared experiences. As Susan Stewart (1993, 133, 135) explains, "Within the development of culture under an exchange economy, the search for authentic experience and, correlatively, the search for the authentic object become critical." She adds,

> We might say that this capacity of objects to serve as traces of authentic experience is, in fact, exemplified by the souvenir. The souvenir distinguishes experiences. We do not need to desire souvenirs of events that are repeatable. Rather we need and desire souvenirs of events that are reportable, events whose materiality has escaped us, events that thereby exist only through the invention of narrative.

The desire for tinfoil must have been partly a desire for authenticity, for what had really transpired. More than any souvenir program or photograph, though, the tinfoil records suggested the authentication of actual sounds that had been shared. If they lacked real readers and real authors, the records were stunning harbingers of provenance. Each had a precise point of origin (recorded at one moment in "real time," according to today's parlance), and they were together defined by virtue of their precision, a new, more vivid indexicality.

Tinfoil offered a new, precise sort of quotation, in effect, or a way of living with the question of quotation as never before. To put it another way, the tinfoil souvenirs suggested that oral productions might be textually embodied as aural reproductions, rather than as the usual sort of graphic representation, spelled out and wedged between quotation marks on a page. What the phonograph demonstrations offered was not the special performance of texts (for example, a written declaration making a nation independent, paper instruments of law making law, or canonical texts making a national tradition). What had been witnessed was a special textualization of performance—quotation somehow made immanent, quotation *marks* of a new sort, which turned (or *re*turned, mechanically, magically) into the quotations themselves.

Tinfoil souvenirs offered the renovation of the souvenir as such, hinting at changes to the then normal connections between matter and event, stuff and utterance, text and speech act. The phonograph demonstrations were indeed only reportable, not repeatable, in Stewart's terms: it was devilishly hard to get a tinfoil record back onto the machine once it had been removed, and certainly impossible to reproduce anything from it when it had been ripped up and distributed among the audience. Still, what was reportable about the demonstrations was precisely repeatability. Narrating the meaning of the tinfoil at breakfast meant testifying to the pending usurpation of that very narrative. These were souvenirs about souvenirness. The desire for authenticity would finally be consummated, when some soon and now imaginable souvenir spoke for itself. The morning papers promised as much. Tinfoil scrap and newsprint squib lying side by side on the breakfast table served to interrogate that promise, if also to prompt the witness/auditor who owned them and formed part of their collective subject. Evidence is scanty, to be sure, and I must admit to having imagined U.S. breakfast table readers in 1878, somewhat in the manner that Benedict Anderson has imagined them imagining themselves as a national community. It should be noted, however, that part of my argument is precisely about that scantiness, about the necessity (then) if speculative (today) reciprocity of newsprint and tinfoil scraps as matters of evidence, where "matters of evidence" and the data of culture are constructed culturally as part of an ongoing dialectic with emergent media forms.

Keyword: Record

Although the exhibitions of 1878 were short-lived, Redpath quickly went on to fresh projects, and souvenir sheets of tinfoil were soon forgotten, the medium of sound recording had forever questioned the relative meanings of writing, print, and public speech. Other versions of related questions soon captured public attention as, for example, when the shooting and eventual death of President James Garfield in 1881 tested the incipient liveness of U.S. media, challenging the press to keep its bulletins up to date via telegraph communications while it followed the hapless use of Bell telephone technology to detect the assassin's bullet inside the president's body.[35] Whether it worked well now or later, recorded sound had disparaged print by implication, helping to suggest the artificiality of writing in comparison to speech. At the same time, the evident inadequacies of tinfoil records as permanent or indelible inscriptions helped to raise emphatic questions of loss within which the meanings of writing and print had long been enrolled. These questions of loss were narrowly a matter of words written down, printed up, and saved. But they were further suggestive of much broader matters of public memory, self-identification and corresponding exclusion, "our Washingtons" and "our Lincolns" saved against "our" uncertain future, for example, and Onondaga and Tuscarora "accents" "preserved" against "their" imminent demise. If their immediate subjects were the local and the semicarnivalesque, the first phonographs were more broadly the instruments of middle-class hegemony, of an Arnoldian "Culture" with its sacralizing functions directed at "our" traditions and its salvage mentality directed at "theirs."[36]

I am suggesting that the phonograph exhibitions formed vernacular experiences of the relationship between speech and writing—a relationship theorized only later by linguists and philosophers—and that such experiences had broad, if unexamined, consequences for cultural formations.[37] These experiences were certainly not unique to the phonograph exhibitions or 1878. They had been and continued to be habitual accessories to print. For instance, in 1869 a congressional committee investigated the treatment of Union prisoners held in the South during the Civil War. The committee canvased the North for testimony, oral and written, with the immediate aim to render those "transient and somewhat *fugitive* histories based on personal experiences and observations" into "an enduring *record,* truthful and authentic and stamped with the national authority." Oral and personal histories were captured and redeemed by the authoritative, textual operations of the incipient welfare state.[38] Diction makes this a particularly tidy example, but so too did the collectors of song, authors of regionalism, critics of idiom, and reporters of news seek to render personal and aural encounters into material, public records. Wherever it appeared, "speaking" on paper was party to precisely the tangle of concerns that Edison's phonograph newly helped adumbrate.

Edison boasted to the newspapers that his invention would ruin the market for books, reasoning that recordings were created naturally at no expense, by sound waves impinging on foil, rather than laboriously set in type by wage-hungry compositors. Authors and their audiences would win, even if printers and compositors would lose. But when the inventor's comments reached the readers of one paper in Philadelphia, set in type of course by compositors, the inventor was quoted as saying that a phonograph record "is not in tpye *[sic]*, but in punctures" on tinfoil. The typographic error is exactly that: graphic. It produces an unpronounceable "utterance," which remains viable only on the page, seen and not said. The four individual pieces of type remain insistently, visually, sorts of type. They are emphatically t, p, y, and e, since they do not make the word *type*. "Pye" or "pie" is a printer's expression for jumbled type, yet there is no way to know for sure whether this was an accident or derision on the part of a compositor employed by the Philadelphia weekly (lazy or Luddite? rushed or radical?), and it is unlikely that many readers stumbled or that any noticed their dilemma.[39] One typo in a local paper is a trivial matter, then, but one that neatly captures the verbal-visual status of writing and print that the first phonograph records visited so unfamiliarly: errors can be quite telling, as chapter 4 will elaborate below. Consuming "speech" on paper requires eyes not ears. By the same token, early auditors wrote "Behold!" (instead of "Listen!" or "Hark!") "The voice of the author is heard repeating his own composition." They got the sensory apparatus wrong in their rush to celebrate the greater immediacy of a new medium within the contexts of its public exhibition.

Far less trivial than the typography of one word by Edison on the future of literature was Moses Coit Tyler's two-volume version of the American literary past. Tyler was one of the first professors of American literature. His *History of American Literature, 1607–1765* (1878) was among the first works to put together a coherent narrative of early American literary history, and it helps characterize the specifically *cultural* status of writing and print at the moment Edison's phonograph was introduced to U.S. audiences. Tyler's volumes assumed as well as redacted a national literary tradition, offering long quotations from early authors.

As he himself described it, Tyler's (1878, xii) project sought to present an exhaustive history of those writings that "have some noteworthy value as literature, and some real significance in the unfolding of the American mind." His method had been to troll for years through libraries, seeking out neglected books, assessing their literary merits, and judging their pertinence to "the scattered voices of the thirteen colonies," which eventually and so importantly, in his view, "blended in one great and resolute utterance" (xi). It was a hugely ambitious project, and one Tyler everywhere portrays in terms that seem to mix the functions of speech and writing. He wanted to see just where "the first lispings of American literature" took their departure from their "splendid parentage, the written speech of England" (11). Tyler's "lisped literature" and "written speech" confirm the degree to which spoken and written language were integrally, mutually, defining. The substitution of "Behold!" for "Listen!" partook of an ancient tradition still current. Writing in general, and literary tradition in particular, lived and was valued in intricate association with speech acts.

The word *record* encapsulates these points. Tyler starts his first chapter proposing the intellectual history of the United States: "It is in written words that this people, from the very beginning, have made the most confidential and explicit record of their minds. It is these written records, therefore, that we shall now search for that record" (5). Despite his reputation as a "superb stylist," Tyler's double use of *record* is confusing.[40] Mental records are his broader category, within which written records stand preeminent as searchable traces. Confusion between the two is indicative of tensions surrounding the potentially national and literary character of textuality for Tyler, and more generally, the gap he seems anxiously to have sensed between minds and pages, between an author's conception and its written, printed expression. Whether or not Tyler composed these sentences just before, after, or in light of Edison's records, his diction partakes of the selfsame context.

Not surprisingly perhaps, Tyler took liberties when quoting the literary tradition he established. Acting on a distinction that would be articulated in copyright law only a year later, Tyler sought and valued authors' ideas, but had to make do with their printed expression.[41] His literary criticism was aimed at works by authors, yet relied on texts by printers. Tyler (1878, xv) noted provenance sloppily, silently modernized spelling and punctuation, and considered that it was "no violation of the integrity of quotation" for him to expunge or correct any confusion, "extreme inaccuracy," or "palpable error of the press." Against such liberal quotation practices and mistrust of the press, and in keeping with his larger project to collect and preserve an explicitly American, explicitly literary tradition, Tyler's *record* is a keyword in the sense that Raymond Williams (1976) picked out his *Keywords.*[42] Like Williams's word *culture* (or *class,* or *media*), that is, Tyler's *record* seems to have "acquired meanings in response to the very changes he proposed to analyze."[43] The term cannot be defined without recourse to the very conditions it connotes: the lispings of literature and the construction of tradition. So too must Edison's "records" have emerged as meaningful according in part to the differently desired subjects of recording. Whether glimpsed here in American literary history or early sound recording, the subjects and the instruments of public memory cannot be pulled apart.[44]

Though obvious in Tyler, this reflexivity of the noun *record* proved fleeting. After some initial awkwardness, its use soon stabilized, broadening to include phonograph recordings as well as one other "curious perversion of meaning" noted by verbal critics. Whereas the word had long meant "an authentic register," including the abstract, immaterial, and impersonal register of public purview, it was now "perverted" to refer to a person's past performance. Originally applied to a candidate's past performance, this sense of the word soon applied colloquially to anyone. People had records. People had *permanent* records. These sorts of records were both personally derived and publicly constituted, possessing a curious instrumentality in the sense that "when a man breaks the RECORD he makes the RECORD" (Farmer 1889, 454).[45] These new records, somewhat like Edison's new ones, were performative. It was as if the frequent personifications of the phonograph had been inverted lexically, as (the meaning of) *record* now incorporated people. Phonographs had so readily become metaphoric authors, readers, and speakers, but people correspondingly became metaphoric machines.

1889–1893: Bottled Bands

Like any new medium, the tinfoil phonograph derived its meanings from both its contexts and public participation. Its immediate contexts included everything from Edison's grand projections of the future and related, recurrent themes in the press, to the formulas of the exhibitions, the collection of tinfoil, and the complex semantics of *record,* recording, and playing records. Broader still were the varied contexts of contemporary electronic media, print, and public culture. New contexts soon emerged, helping to recast the new medium, and the habits of public participation also changed, along with the technical and economic structures that pertained to them. Nickel-in-the-slot or "automatic" phonographs replaced exhibitors and their machines; audiences no longer recorded themselves; and the rational (albeit semicarnivalized) technocracy of the lecture hall gave way to more commercial and casual public encounters with popular music. In the place of tinfoil sheets, the nickel-in-the-slot machines used wax cylinder records and offered no tangible souvenir.

Though their success was short-lived, nickel-in-the-slot phonographs have long been seen as an important transition point in the history of the medium. They taught capitalists and musicians alike that phonographs made sense as amusement devices to play prerecorded musical selections—something that audiences (and the inventor Emile Berliner) somehow seemed to have understood a few months or years before. While this characterization is an accurate one, it does little to acknowledge the nickel-in-the-slot machines on their own terms. As Jonathan Sterne (2003, 203) notes, "If we consider early sound-recording devices in their contemporary milieu, the telos toward mass production of

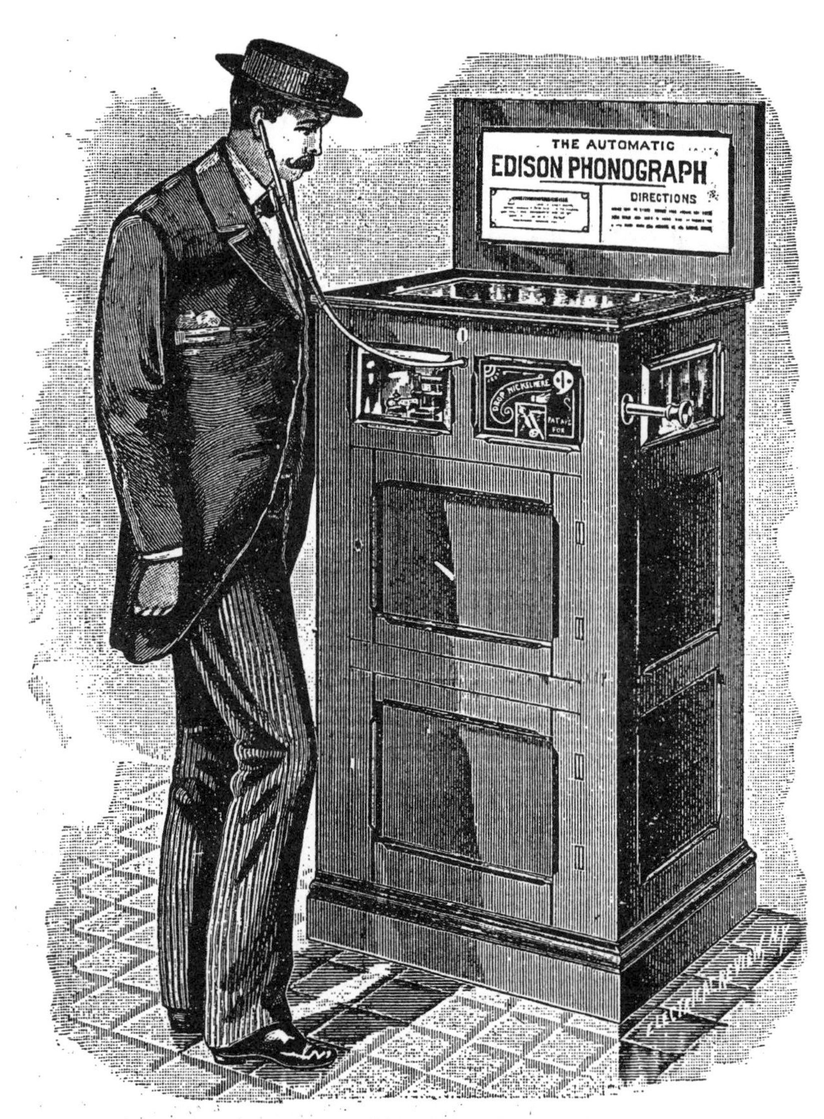

Figure 1.2 Nickel-in-the-slot phonograph from *The Phonogram*. 1892. (*Source:* Edison National Historic Site, National Park Service.)

prepackaged recordings appears as only one of many possible futures."[46] Neither the promoters nor the patrons of the automatic phonographs could read the future, and their experiences of what the nickel-in-the-slot machines meant must be plumbed in different ways. One part of what they meant had of course to do with writing and print, with the older, long-held meaning of "records" as inscribed and factual embodiments of data important or potentially important to an abstract public. Phonograph records, even musical ones, must have first made sense according to the ways they adapted the intuitive facticity and publicity of written records, as quickly as this adaptation must have been forgotten, as the new meaning of *record* lost its unfamiliar breadth, and as the new medium gained more and more formal and functional conventionality.

The nickel-in-the-slot phonograph was invented by Louis Glass, the general manager of the Pacific Phonograph Company, which was one of thirty-two "local companies" organized between 1888 and 1890 to exploit Edison's "perfected" phonograph (and the related graphophone) in protected territories across the United States. Edison's improved phonograph may have used cylindrical records instead of sheets, but its purpose was still primarily textual; it was a business machine for taking dictation. Berliner had already imagined using his gramophone, first demonstrated publicly in 1888, to play prerecorded musical "phonautogram" discs, but the Pacific Phonograph Company and the other local companies would be slow to see amusement as anything but a sideline. Glass's first slot machine—then three, of them and soon fifteen—proved phenomenally popular with patrons in San Francisco, and by the end of 1890, his licensees and competing devices were playing prerecorded cylinder records in public "parlors," saloons, hotels, depots, drug stores, and arcades throughout the United States.[47]

Like the first phonographs with their sheets, the nickel-in-the-slot phonographs had structural as well as functional properties crucial to their definition. Social context and musical content both played significant roles in the meanings of these new machines, but so did the shape of the technology itself. Despite their name, the nickel-in-the-slot machines possessed at least three design attributes of equal importance to their coin-activated play mechanisms. For one thing, they each had to contain a return device, so that when the reproducer (the mechanical part that "reproduced" the sound) reached the end of the record, it could return to the beginning and be reset for the next coin. Repetition was built into the machinery. Each patron's deposit of a coin formed a single transaction among an implied infinitude of transactions that were exactly and *automatically* the same, although the earliest slot machines were easy to fool. Early "profits" were frequently reaped in slugs, and reportedly "even bits of ice from saloon patrons' drinks" could be used to activate the play/return mechanisms.[48] Like other vending machines of the period, the nickel-in-the-slot phonographs helped standardize and depersonalize transactions, even if they briefly helped question or make risible a standardized nickel. The phonographs joined the first few "silent salesmen" in the urban United States, emphasizing by contrast the muteness of automatic gum machines, stamp machines, and scales (Schreiber 1961, 12–20).

Two further design attributes were less imperatives than they were matters of choice by designers. While the nickel (or "nickel") disappeared from view into its slot, and the wet cell battery powering the phonograph was safely out of sight below, the record, reproducer, and return mechanism were usually visible to patrons, often located under a glass window or dome at the top of the machine. In this way, the performance of these machines was public, while the (electrical and financial) power behind their performance was private and mystified in a bit of oak cabinetwork. Finally, unlike later jukeboxes but like dictation phonographs, the slot machine phonographs usually played for one patron at a time. The modest volume of early recordings made it preferable to use headphones or "hearing tubes" in connection with the machine. These resembled the biaural stethoscope, a diagnostic tool just coming into use at the time (Reiser 1978, 40–43). Customers privately heard a record in somewhat the same way that a doctor heard a patient's lungs. Recorded sounds described in 1878 as faint and "far-off" now resonated immediately in a person's ears. "So distinct are the sounds," one account put it, "that you imagine that every one within a radius of fifty feet must hear them," but all they heard was "a faint mum, mum, mum," even if they could see through the glass.[49] Users paid for private, even intimate encounters with public machines. Something like the necessity of watching projected motion pictures in the dark just a few years later, coin-operated phonograph parlors disaggregated the senses, helping to divide customers from one another even as they drew them into anonymous crowds. Parlor patrons stood together, saw together, but listened by themselves.

Even phonograph insiders admitted that it was preferable to have an attendant, "operator," or "inspector" available to keep nickel-in-the-slot phonographs in good working order, change musical selections every so often, and wash the ear tubes "every few days." Slot machines were stationary devices, placed by companies in advantageous locales, and if an inspector tended more than one machine or group of machines, it was the inspector who circulated. Not so with a related form of exhibition phonograph, which was traveled around to fairs, picnics, resorts, and small towns by itinerant showmen. These exhibition machines boasted eleven or "fourteen- and sixteen-way hearing tubes," using what were termed "way rails" in the trade to disperse the sound among eleven to sixteen different sets of headphones at once. Patrons paid their fee, which was collected by hand, and then stood like so many doctors in consultation over a single patient (or like veterinarians over a single animal, since in one surviving photograph, the exhibition phonograph is mounted on the back of a donkey). They listened in a group, though their intragroup communication was disabled by each individual's act of listening; they heard the same thing in the same "way," but the hearing tubes divided them.[50]

I am suggesting that the design of nickel-in-the-slot and exhibition machines helped to create vastly intricate experiences of public and private—experiences animated by distinctions between performance and power, seeing and hearing, dead matter and living voices. These were repeated, repeatable experiences that suggestively tended to standardize and depersonalize exchange, to collect and yet atomize consumption, and thus effectively to essentialize the marketplace, making it more easily experienced as an abstraction: the market. These experiences simultaneously informed and were informed by the social, largely urban contexts of the nickel-in-the-slot phonographs as well as the sound content of the records that they played.

The nickel-in-the-slot phonograph helped to negotiate what historian Kathy Peiss (1986, 6) calls "the shift from homosocial to heterosocial culture" in the United States, and did so in large measure because of the range of differently public venues it played.[51] The Columbia Phonograph Company was the most successful of the local phonograph companies, able to place many dictation phonographs and take advantage of the nickel-in-the-slot trade by putting 140 machines in its Washington–Baltimore territory by the end of 1891. The company apparently targeted "*all* places where many people congregate," yet its nickel-in-the-slot phonographs were concentrated in groups, at hotels, train stations, and drug stores.[52] The first of these urban audiences must have been diffuse and masculine. Hotels and depots were long-standing communication nodes, where telegraph intelligences and news were available to a mobile and disproportionately male audience. By contrast, drug stores, summer resorts, and amusement arcades offered a decidedly heterosocial setting for recorded sound. The Palace Drug Store sported "the finest fountain and the best soda trade" in New Orleans as well as at least one extraordinarily profitable nickel-in-the-slot phonograph placed by the Louisiana Phonograph Company—a phonograph that served as both "an attraction and amusement for patrons." Groups of machines were sometimes gathered into "parlors," their sponsors enticing female and male patronage by referring paradoxically to this public space as a kind of middle-class domestic sanctum, in the same way that the railroads ran their "parlor cars." It was a designation that may have helped ease the later progress of these amusements into U.S. homes, where they joined such parlor media as stereoscopes, and where they presaged radios and televisions in the "living rooms" of the next century.[53]

Saloon patrons, male and working class, got to try the phonographs too, and like so much other technology, the nickel-in-the-slot machines possessed resilient masculine associations at the same time that they engaged and helped to construct a heterosocial public. One observer described phonograph patrons as "blasé men about town," who prematurely thought they had "squeezed all the juice out of the New York [City] amusement lemon," but now went around to hear "a little bit of the very latest things." In Buffalo, automatic phonographs eventually had to be discontinued, because saloon patrons amused themselves by "beating" the machines, by tying a string to a nickel, for instance, and pulling it back out of the slot. Bartenders were supposed to receive a commission to oversee the machines and prevent such countercapitalist sociability, but as one of them put it, "The commission went to the bosses and we were not at all interested. See?"[54] Unlike the frequency and regularity of news accounts that rendered phonograph exhibitions during the summer of 1878, the technology that inspired "squeezing" amusement and "beating" machines as forms of class-aligned male sociability in the early 1890s appears only occasionally and irregularly in the press. As Mary P. Ryan notes, "The public record seldom detailed the everyday life that transpired in the smaller nooks and crannies of urban space" (1997, 209), no matter how complicated the "alloy" (206) of public and private concerns that informed places like the barroom, the corner grocery, and the stoop.

If the "Behold!" of 1878 relied on the reason and "open" disinterestedness of the public sphere, the bartender's "See?" of 1892 hinted at the clash of interests that muddied everyday life. Continuing in the tradition of newspaper wit, one St. Louis paper included a fanciful barroom encounter under the headline "Automatic Robbers":

> "That's right, sir. Give it a punch about an inch under the slot. That sort of rattles the nickel into place. You see the incline isn't steep enough to carry the coin down to the machine. By jarring it up you get the machine started. Don't you see?"
>
> "See," yelled the man with the tubes tucked in his ear, "Good God, man, I don't want to see, I want to hear."
>
> The man who is doing the explaining is the bar-tender at the St. James. The man who is swearing is the victim of a phonographic holdup game.[55]

Here, the unfamiliar idiosyncrasy of machinery, the disaggregation of senses, and the hucksterism of commerce all helped to define the sound of the phonograph and its recording. (Even the managers of the local phonograph companies admitted guiltily that their machines worked poorly, confessing "that the people who pay their money" should really "get something for it.")[56]

What did people hear when they dropped a nickel in the slot and put on hearing tubes? The first thing patrons heard was the medium itself, the whir of a motor and the scratch of a reproducer point against wax, then there was a quick announcement, and a recorded performance lasting about two minutes. Like the experience of listening to tinfoil records, the experience of listening to a nickel-in-the-slot phonograph is difficult to recover, largely because the issue of mimesis is so vexed in hindsight. It is impossible to gauge precisely the extent to which listening to a record of ♫ was experienced as ♫ or a performance of ♫, for instance, and how much listening to such a record was an experience of mimesis, of listening to a *representation* of ♫ or its performance, with the many and complicated parameters available to representational forms: material, authorial, semiotic, and so on. Was the phonograph making music, or was it (just?) playing music? To what extent was mechanical reproduction a transparent re-production? While it can readily be supposed that listening to a record of ♫ was experienced as ♫ *and* as a representation of ♫, the balance and interplay between the coeval alternatives, real and representational and performative, must have been a purchase of the moment, a matter of locality as well as shared cultural practice, and can hardly be accounted for completely by studies of that later construction, "acoustic fidelity."[57]

What can be known for sure about the experience of listening to nickel-in-the-slot phonographs is something of the recording artists, their repertoire, and the conventions of their recorded performances. Despite growing evidence that they were mistaken, the executives in charge of the various local phonograph companies had a hard time weaning themselves from the idea that automatic phonographs should play recorded speech instead of recorded music. Speech was a "greater marvel" than music, particularly instrumental music, since so many had heard things like music boxes before. And speech was "good." As one man said of audiences in the Midwest, "I think that good talk on the machine, to hundreds of people in these smaller places, is much better than rotten music; it is much more of an entertainment to them—some sort of brief address or some short story."[58] Goodness thus came larded with assumptions about class, taste, and provincialism as well as the poor sound quality and relatively lowbrow fare of the available records.

The phonograph companies produced recordings and offered them to each other for sale in printed trade catalogs, the first in 1889 and 1890. A single typed sheet by the Columbia Phonograph Company lists sixty available recordings, "Selections Played by the U.S. Marine Band of Washington, D.C." A printed sheet issued by the parent North American Phonograph Company lists fifty-five selections divided into "Brass Band" numbers (sixteen of them), "Parlor Orchestra" numbers (fifteen), "Cornet" pieces (sixteen), and "Clarinet" pieces (eight); in this case, no recording artists are named. In both catalogs generic categories like march, waltz, and polka appear as prominent data, as a means of organizing selections, in the titles of individual selections, or as explanatory notes accompanying selection titles. Record catalogs from the next few years follow a similar pattern, indicating the importance of genre and instrumentation (or vocal range) to the identification of records, often at the expense of named recording artists, and attesting to the predominant production of recorded band music.[59] By 1892, the Edison Phonograph Works was offering 143 different records, still heavily weighted toward band selections by the U.S. Marine Band and others (forty-six in all), along with many vocal and other instrumental records, a little artistic whistling (four), a few recitations (three), and a couple of the first "Darkey Songs." The Ohio Phonograph Company issued a similar list, and the Louisiana Phonograph Company offered a minstrel stump speech among its band and other records.[60]

The repertoire of these first recordings has been explained by some as the de facto result of an imperfect technology: whistling recorded well, so did tenors, but sopranos did not. Whatever the merits of this reasoning, it seems clear that the early predominance of U.S. Marine Band records had much to do with the early dominance of the Columbia Phonograph Company, which was tapping its local talent in Washington, DC, when it recorded "the President's Own" band. Columbia supplied other companies with records, which it produced five at a time, and it used the courts to block others from recording the Marine Band. As one newspaper report put it, the Marine Band was "render[ing] itself immortal . . . by having its most harmonious strains bottled in large quantities." Band members played to an empty "room on E Street below Seventh," but "for the entertainment of people in all parts of the United States."[61]

Records of the U.S. Marine Band, like the U.S. Marine Band itself, articulated connections across local and national communities. The band was founded in 1798, moved to Washington, DC, in 1800 with the federal government, and became known as the President's Own during the Jefferson presidency. It regularly played at the White House (as it still does) for state functions and came to share the transcendent localism of Washington, DC, as a national symbol, although it also toured throughout the United States.[62] The availability of Marine Band recordings added transcendence of another sort, suggesting one way in which the private hearings of individual nickel-in-the-slot patrons must have possessed a national and even nationalist valence. Selections played and recorded by the Marine Band were not all "Semper Fidelis," "Star Spangled Banner," and "Red, White, and Blue," of course. For instance, the band recorded a "Kaiser Joseph" march, a "Sweetheart Waltz," a "Mexican Dance," and the immensely popular "Irish" song "Down Went McGinty" in 1890. So sonic representations of ethnic, racial, and national difference were already helping the medium in its inscription of "American" interests.

Any band music played by any band must have possessed a different and yet related set of associations for nickel-in-the-slot auditors. Band music formed part of the experience of public and local spaces in the United States, as the accounts of Kenneth Kreitner (1990) and Margaret Hazen and Robert Hazen (1987) make clear.[63] There were an estimated 10,000 bands in the United States in 1889, and among them perhaps 150,000 band members.[64] Towns with populations as small as two thousand sported amateur bands, identified civically and with ethnic or trade groups. They were usually named for their towns or sometimes a bandleader known about town. Their members were generally lower- to middle-class men, and band membership fluctuated to the extent that this remained a relatively itinerant class of workers and tradespeople. Instrumentation, uniforms, and talent varied widely, as did repertoire and musical arranging, as far as can be determined. The bands played a variety of venues and events, including outdoor concerts, benefits for themselves or other groups, and special events and holiday celebrations. They played when a new telephone line went in or the lecture circuit brought celebrities to town.[65] They marched through the streets, and performed from bandstands and standing or sitting where needed. They played as often as every week in the summer and maybe not at all in the winter.[66]

The bands were in some sense a substitute for and the ultimate successors to an earlier parade tradition. If their personnel suggests a voluntarist, male citizenry performing acts of public representation, the sounds of their play must have powerfully articulated a local community, indicative of common and public interests while doubtless helping to perform gestures of inclusion and exclusion based on differences of gender, class, race, political partisanship, and nativism.[67] Though cities had plenty of bands, of course, band music tended to connote the auditory landscape of more rural life. John Philip Sousa (1906, 281) lived in Washington, DC, but he rhapsodized about "the village band" and "the country band, with its energetic renditions, its loyal support by local merchants, its benefit concerts, band wagon, gay uniforms, state tournaments, and attendant pride and gayety." Like Old World village bells, village bands in the United States produced auditory markers of identity, dense with messages about home and leisure, if possibly too with messages about nation, state, and civic ceremony.[68]

The nickel-in-the-slot phonograph brought this outdoor, public music inside, "bottling" and saving it under glass for every person's public tender and private audition. Patron's each heard their own band for the first time: they together had individual, individually purchased experiences of common, communal sounds. Those sounds were the sounds of public life, sounds for the first time disembodied. Like the eighteenth-century print artifacts Michael Warner has described, in other words, these brassy strains were themselves "metonyms for an abstract public," newly "embodied" in cylinders of wax.[69] This is to suggest that these early band records, because of the public and civic contexts of band music, and the complicated spatial qualities of sound itself, offered an additional complication to the ways in which nickel-in-the-slot phonographs helped to rearticulate private and public.

Nor was the consumption of nickel-in-the-slot recordings uninvolved in the relative meanings of print and public speech as the substance of public record. Though musical performances increasingly formed its intended subject, the emergent medium was far from transparent in its delivery of that content. Each nickel-in-the-slot machine appeared under a card—the companies often called them "announcements"—which gave proprietary information, instructions for operation, and indexical information about the recording it was prepared to play. Patrons were all accustomed to the consumption of print like this in public. Recorded sound existed, that is, within public space already profoundly self-iterated by the printed word, an unrelenting profusion of commercial messages, and plenty of noncommercial signage with the abstract public authority of "Keep off the Grass" and "Post No Bills."[70] Patrons read announcements, and they might then play records, watching as a reproducer moved slowly across the surface of each recording, returning back the other way for the next play. The phonograph's early identification as a textual device and the slow, lateral scanning of the reproducer during play may have helped its audible content make sense in the context of public print consumption.

Despite the terminology of "announcements," the phonograph companies knew the rhetorical value of the printed cards as advertisements. The manager of the Ohio Phonograph Company urged that if records appeared with perfunctory announcements of "what the man may hear," they did not do very well, but "if you will put on the full announcement, stating what it is, in as effectual a way as the circumstances will warrant, you will observe an increase in receipts." He himself had gotten "an inferior cylinder" to pay handsomely with this sort of rhetorical frame; a tired old banjo record announced as "an-old-time-before-the-war banjo song sung by a plantation [slave]" made $4.75 a day, "away ahead of the Marine Band receipts."[71]

Variously framed by print and privately consumed through ear tubes, the nickel-in-the-slot records adapted yet another mode of public address in the quick announcements that prefaced recorded performances. All recordings began with shouted announcements of this sort: "The following record taken for the Columbia Phonograph Company of Washington, D.C., entitled 'The National Fencibles March,' as played by the United States Marine Band"; or "'The Bowery,' from 'A Trip to Chinatown,' as sung by Mr. John Yorke AtLee for the Columbia Phonograph Company of Washington, D.C., accompanied on piano by Professor Gaisberg."[72] Different announcements apparently rendered different kinds of indexical information, but they all gave the title of the piece performed, and many said something of the recording artist or artists. All announcements contained proprietary information to identify the record production company as an interested party. These spoken announcements worked in tandem with the printed advertisements and instructions, framing the selection that followed.

The recorded announcements helped to construct the intricate performative qualities of the recorded selections, which were rendered "as played by," "as sung by," or "accompanied by." Recorded announcements located recordings in a performed past tense, but did so in their own, abstract, and continuous present of some patron plus some nickel. They spoke with an impersonal authority that was neither the recording artist's nor the company's, but that put the listener in a consuming relation to both. Nothing separated them as recordings from the recorded musical selection that followed except their different or differently intentional voice. (In the second example above, the record was both announced and performed by AtLee, who was a moonlighting government clerk in Washington, DC.) They announced a recording "by" and "for" someone, but were themselves neither explicitly by nor for anyone in particular. If band records offered metonyms for an abstract public, the recorded announcements offered metonyms for an abstract authority. Gone were the fumbling professor and the entrepreneurial showman, replaced instead by the male, stentorian, disembodied voice and indeterminate public address of media administration, hailing its subjects: like those later messages, "We interrupt our broadcast to bring you these important messages," "Stay tuned for scenes from next week," and "File not found."

Without sources to suggest audience response, one can only wonder at the extraordinarily complicated framing functions such announcements must have subtly performed for the small number of "topical" recordings proffered by the Columbia Phonograph Company. Columbia offered a few recorded campaign songs for the 1892 presidential election (Harrison versus Cleveland), a proletarian quartet (judging from its title), "The Fight for Home and Honor (Homestead, Pa.)," and the stunningly self-referential and self-fulfilled "You Drop a Nickel, We Do the Rest." Patrons would hear "You Drop a Nickel" announced after they had read the title and followed its instructions to drop a nickel. The printed and recorded announcements were at once upstaged and enacted (and undercut, if a counterfeit nickel were involved). Then an abstract, unspecified "we"—represented by the announcer's singular voice—began, continued, or repeated to "do the rest" in playing a past performance by specified others (AtLee and Gaisberg again, I think). The machine and patron performed an elaborate ritual of mimetic interaction, a *seeming* dialogue of representations more or less significant to the momentary location of "the rest" as an experience, a commodity, and itself a representation. Of moment was likely the Kodak Company's familiar slogan, "You push a button, we do the rest," which enrolled the same characters of "you" and "we" into the consumption and production of lifelike representations as commodities.

In place of the preservative or memorial qualities of tinfoil souvenirs, nickel-in-the-slot phonographs offered fleeting experiences of wax and nickels. Newsprint and tinfoil had offered ample contexts for one another, but wax and nickels only rarely got covered in the press. During the summer of 1878, similar accounts of the phonograph cropped up in local papers everywhere, but accounts of the automatic machines of 1889–1893 were fewer and are less formulaic. This may have been true partly because the dissemination of automatic phonographs was less coordinated, attempted for a longer period, and more singularly urban than the wave of tinfoil demonstrations in 1878, and partly, as I have hinted above, because their functions as "cheap amusements" and barroom furnishings trivialized them as subjects. (My small archive is mostly Edison's, since the inventor hired clipping services to collect public accounts into his private files.) For whatever reasons, and although nickel-in-the-slot phonographs clearly, intricately, helped to reconstruct publics in relation to private consumption, they were not themselves important "matters of record." At least, that is, until media historians—knowing what happened next—started to tell their story as a crucial turn on the road to mass media, as harbingers of what gets called "the industry" today. Though more significantly matters of public record in their day, tinfoil phonographs have tended to drop out of media history for the same reason. Looking ahead at what happened next, the demonstrations of 1878 don't make a lot of sense, and tinfoil phonographs get only perfunctorily or "curiously" noted.[73]

Though later mass media contexts have thus tended to reshape their meanings retrospectively, neither tinfoil phonographs nor nickel-in-the-slot machines can themselves be confused as mass media. They were new media, but not yet "mass" either in their scale or political economy. First, simply in terms of scale, their reach was modest. Between May and October 1878, for instance, the Edison Speaking Phonograph Company paid Edison his exhibition royalties in the sum of $1,032, suggesting that some 82,553 people had paid a quarter to see the show, although there was evidently plenty of trade before May, after October, and off the books.[74] Likewise, in 1891 the local phonograph companies had only 1,249 nickel-in-the-slot machines in operation around the country, less than half the number of dictation phonographs they had rented out to offices in the same year. Seventeen of the local companies voted at their convention that the amusement machines were profitable, while two companies voted "no." While 1892 was a better year, the economy slumped dramatically again in 1893.[75] In short, more people in 1878 certainly read about tinfoil phonographs than saw or heard them, and more people in 1889–1893 listened to live band music than heard it played on nickel-in-the-slot machines.

Despite such modest proportions, the earliest history of recorded sound offers some interesting lessons about the emergence of new media and about media as the subjects of history. Among the most obvious lessons is the failure of the "beta" device unveiled to public acclaim to presage anything like the functions that subsequent, related devices eventually serve. That the social meanings of new media are not technologically determined in any broad sense should be clear. The technology in this case proved to be fertile ground for reinterpretation. What happened, as Jacques Attali (1985, 89) says, was "a massive deviation from the initial idea of the men who invented recording." Whereas "they intended it as a surface for the preservation" of speech, what they got was a host of new cultural formations: new social practices for producing and consuming music, new corporate structures for capitalizing and disseminating performance. These new practices and structures mutually entailed a new mass medium, where the connotations of "massive" deviation and "mass" media go well beyond questions of scale, scope, or Attali's surface to point instead toward a substantially new organization of publics in relation to markets, and consequently, an emerging identity between citizens and consumers. This new organization or "massification" forms the subject of the next chapter, which addresses the broad social contexts within which recorded sound became sensible and then intuitive as a medium for playing prerecorded music in the home.

Obviously, though, the lessons of the past need no narrow subscription to the present. The earliest history of recorded sound points broadly to the coevolution of new media and media publics. The prephonograph media public in the United States was characterized by an increasingly diverse and mobile population, and by the dominance of print forms that were increasingly numerous and may have been increasingly uncertain in their significance for readers. As such, audiences experienced and helped to construct the logic of recorded sound, responding to specific material features of the new medium as well as the changing contexts of its introduction. Relevant material features included things like tinfoil sheets and rubber hearing tubes, while relevant contexts ran the gamut from lyceum halls and barrooms to the existing organization of the local and national press, from the immediate circumstances of keeping tinfoil souvenirs to the abstractions of the public sphere, which located the kinds of things worth keeping at all. Phonograph records—by definition—became implicated in the discernment of "the public record," and the concomitant matters of record and evidence, even as their very definition remained in flux.

2 New Media Users

The phonograph was one of those rare, Jekyll-and-Hyde devices that was invented for one thing and ended up doing something completely different. Edison "perfected" his phonograph (so he said) in 1888, and was thoroughly convinced that its primary function would be in business communications. His machine had read-write capabilities, and he and successive groups of enthusiastic investors thought it would make a revolutionary dictation device. They were wrong, of course. In the mid-1890s, consumer demand helped to transform the phonograph into a read-only amusement device, and by 1910 recorded sound had become the first nonprint mass medium. The purpose of the present chapter is to account for as well as describe this diversion of purpose, and in doing so, urge that the histories of new media be sought amid uses and users, rather than simply amid descriptions of product development, product placement, business models, or calculations of market share. Those elements are crucial, but that's not all there is. Like the preceding chapter, the chronological focus of this one is narrow, while its ambitions to contextualize new media remain extremely broad. During the years 1895–1910, recorded sound was reconceived as a commodity for home consumption. Somewhat like the much later reorientation of computing toward "personal" computers, the success of home phonographs and prerecorded phonograph records relied in part on unacknowledged assumptions about what were personal and domestic concerns at the same time that it signaled profound changes attending U.S. culture at large.[1]

The title of this chapter, "New Media Users," echoes and adapts the title of the previous one, "New Media Publics," in part because I want to notice distinctions between publics and users that are too often forgotten or ignored. Publics are comprised of users, but not all users are entitled or constitutive members of the public sphere. The Indians, shrews, and minstrels who came up in accounts of tinfoil phonographs during 1878—not to mention crying babies, barnyard animals, and inebriates—were neither users nor publics; they were *representations* that differently served to define public life and public memory according to long-standing, if unspoken, rules about who matters and who doesn't, and by what means and media. By contrast, the working-class saloon patrons who amused themselves by cheating nickel-in-the-slot phonographs were true users, but they were hardly the public imagined by phonograph executives, required by capitalist exchange, or comfortably accommodated by arbiters of U.S. public life or middle-class culture during the 1890s. They were hackers.[2] Nor were more dutiful, paying customers all equally members of the public sphere, although one feature of mass culture as it continued to emerge in this period was precisely the apparent eclipse of such distinctions, as publics came gradually to seem comprised of consumers rather than citizens—in other words, as consumer choice came gradually to seem the most effective and available public expression of an individual's reason and identity.[3]

I am being more critical than I am cynical: I want to notice distinctions between publics and users because I argue here that while new media help mutually to reconstruct public life and public memory, it is users who help to define new media in crucial ways. Or as Janet Abbate (1994, 4) puts it in reference to the early history of the Internet, "Users are not necessarily just 'consumers' of a technology but can take an active part in defining its features." Users in this sense do not necessarily stand in any self-conscious relationship to publics. They are neither exactly "counterpublics" nor exclusively subcultures; they are diverse, dynamic, and disaggregate. They stand both as mirrors and receptors for the ideological formations of the public sphere, yet are not themselves necessarily ideological: individuals do not "belong" as users, but their activities as users can have profound consequences for what Michael Warner (2002, 12) calls the "metapragmatics" of belonging.[4] Of course, the users of early recorded sound were not active in exactly the same manner that users of Abbate's early Internet were, but they helped to shape the medium in important ways. In particular, gender difference became integral to the definition of recorded sound. Middle-class women were central to the meanings of phonographs and records as such because women helped deeply to determine the function and functional contexts of recording and playback. Put simply, I do not propose that home phonographs eventually became gendered instruments of mass culture. They did, but there's much more to it than that: I propose instead that gender and cultural differences were built in to home phonographs from the start.

Somewhat like Abbate in her work on the ARPANET and Internet, my interest is in posing questions that might bedevil the strict dichotomy of production and consumption, which is so familiar to media history and so characteristic of U.S. attitudes toward technology, whether those attitudes are technophobic or technophilic. The production/consumption dichotomy harbors a particular determinism since it typically puts producers first and then draws an arrow toward consumers. Within this gesture lurks a tendency to use invention and technology as sufficient explanations of social and cultural change, and this in turn has helped orient media history toward narratives of social effects, and at the same time away from the agencies of any but white, middle-class men and the developed world. It favors publics over users, in other words, rendering a history in which, for example, "inventing the telephone is manly; talking on it is womanly."[5] A more calculated version of the same logic can lurk behind even the most affirmative, feminist-friendly accounts of consumer resistance as well as celebrations of exceptionalism, rendering a history in which, for instance, men invented the telephone, but women taught them what it was for, or in which men invented the phonograph, but let women help them sell it during the First World War.[6] Such narratives describe women as agents, whether through their adaptive reuse of consumer goods or occasional usurpation of supposedly male roles. Yet their agency is largely reactive rather than active, a bind that can only be undone by a critical reevaluation of production and consumption as either historically stable or mutually distinct terms of analysis.[7]

Habitual reliance on the production/consumption dichotomy has led to an early history of recorded sound that runs something like this: After Edison invented the phonograph, competition arrived from Berliner (the "gramophone") and inventors at Bell's Volta Laboratory (the "graphophone"), prompting Edison's own commercial development of his machine. The phonograph and graphophone were marketed by the North American Phonograph Company, incorporated in 1888, via a network of local companies operating in protected territories. The expensive devices were leased and later sold as dictating machines, without much success, since office workers resisted the complicated and still temperamental machinery. "Almost by accident," things changed: one California entrepreneur adapted his phonographs into nickel-in-the-slot machines, which gradually both proved the success of recordings as amusements and created a demand for prerecorded musical records. When Berliner started to market his gramophone and disc-shaped records in the United States in 1894, he faced competition from imitators as well as companies like the Columbia Phonograph Company and, in 1896, Edison's National Phonograph Company, both of which sold only cylinder records at first. The market for home machines was created through technological innovation and pricing: phonographs, gramophones, and graphophones were cleverly adapted to run by spring motors (you wound them up), rather than messy batteries or treadle mechanisms, while musical records were cleverly adapted to reproduce loudly through a horn attachment. The cheap home machines sold as the ten dollar Eagle graphophone and the forty dollar (later thirty dollar) Home phonograph in 1896, the twenty dollar Zon-o-phone in 1898, the three dollar Victor Toy in 1900, and so on. Records sold because their fidelity improved, mass production processes were quickly developed and exploited, advertising worked, and prices dropped from one and two dollars to around thirty-five cents.[8]

What's missing? Besides the elision of consumption and buying (phonographs and records are *played,* after all), such accounts limit the definition of production to the activities of inventors and entrepreneurs. What if that kind of production were only a tiny part of the story, granted its singular importance by the same cultural norms and expectations that construe technology as a male realm? The very meaning of technology might be at stake. The spring-motor phonograph worked in homes around the world, but would it have been described or even understood as "working" if it did not already make sense somehow within the social contexts of its innovation? For that matter, would the nickel-in-the-slot phonograph have worked in just the way it did if the office workers disparaged as "nickel-in-the-slot stenographers" by the North American Phonograph Company executives had embraced rather than resisted the dictation machine?[9]

Questions like these get women and other users back into history. "Recorded sound," writes historian Andre Millard (1995, 1), "is surely one of the great conveniences of modern life." Yet we know from Ruth Schwartz Cowan's important *More Work for Mother* (1983) and a few other feminist histories of technology just how vested the definition of "convenience" can be within the gendered, social, and economic constructs of a time and a place. Electric washing machines are modern conveniences, but their appeal and adoption at the beginning of the twentieth century in the United States must be read at least against simultaneous changes to the standards of acceptable cleanliness, the demographics of domestic labor, the socioeconomic geography of electrification, and the ongoing reconstruction of other domestic chores.[10] Convenience by itself explains nothing. In short—to bend the language of production just a little—it must be that homemakers helped *make* home phonographs, to the curious and complicated extent that they "made" homes, once it is acknowledged that the lives of new media are not just public relations events, business models, or corporate strategies but fully social practices. As Jonathan Sterne (2003, 197) puts it, the early history of the phonograph "is at least as much about the changing home and working lives of the middle class as it is about corporate planning and experimentation."

I am suggesting that phonographs and phonograph records had rich symbolic careers, that they acquired and possessed meanings in the circumstances of their apprehension and use, and that those meanings, many and changeable, arose in relation to the social lives of people and of tangible things. Perhaps because they are media in addition to being technologies and commodities, phonographs and records seem to have possessed an extraordinary "interpretive flexibility," a range of available meanings wherein neither their inventor nor the reigning authorities on music possessed any special authorial status.[11] Edison's intention for the machine was largely confounded, while composers and musical publications left the phonograph virtually unnoticed until its immense popularity forced them into addressing its role as a "self-playing" musical instrument. Later, hugely influential record producers—eventually multinational corporations known as "labels"—habitually failed to predict which recordings would succeed and which would fail. Instead, the medium of recorded sound was authored by the conditions of its use, with phonographs and phonograph records acquiring their cultural heft as they acquired their range and circulation among human hands and ears as well as among other media and goods.[12]

Though frequently ignored by cultural theorists and cultural historians who tend to emphasize the *extensive* qualities of mass culture, phonographs and phonograph records, like music itself, suggestively exhibited *intensive* qualities to accompany those extensive ones. This extensive/intensive dichotomy is a helpful heuristic that has emerged from the histories of reading practices to distinguish modern and premodern literacies.[13] Simply put, either readers consume a lot of material, moving quickly from one text to another, or they consume a little material repeatedly and with greater intensity. Newspapers and pulp fiction versus the Bible. Modern mass culture involves consumption that is extensive in this sense. Many different media bombard audiences while commodities jostle and vie for their attention. Television viewers flip through hundreds of channels without watching one. Texts and other goods are consumed in a seemingly endless stream, whether the unceasing flows of the broadcast media or the proliferation and perishability of so many mass-produced consumer goods. Mass society is disposable, profligate.

Like other mass media, phonographs and records came to possess extensive, mass appeal, and notably to rely on the consumption of public taste as such. They relied, that is, partly on the marketable extent of their appeal in the marketplace—a consumer logic that lies behind fads, hits, and stars. But phonographs and records also made sense according to intensive uses, at first by customers at public phonograph parlors and later by listeners at home. The present chapter begins by introducing this intensity in a comparison between phonograph records and another contemporary medium, the mass-circulation monthly magazine, which is seen by some critics as the cardinal form of U.S. mass culture, at least before the nickelodeon. The chapter then addresses the definition of the phonograph as a form of mechanical reproduction as well as a musical instrument dependent on women as agents and subjects, where subjectivities arise in part according to the lived experience or negotiation of a host of cultural categories, like professional and amateur, or at home and in public. And the chapter concludes by alluding to the ways in which the norms and habits of shopping helped to define the home phonograph amid the desirability and circulation of other goods.

By sticking to the verb *define* rather than *produce* and *consume,* I want to describe the emergence of home phonographs as a multifaceted, culturally and historically specific process that involved a wide range of factors that were determined in part by the growth of the middle class and changing roles for women in U.S. society. The evidence of such a multifaceted process of "design and domestication" is of course multidimensional.[14] It includes representations of different kinds, like the language and metaphors used to describe phonographs, the advertising used to sell them, and features of phonograph design. And it includes a variety of related social practices, like music making as a domestic pursuit, mimicry as a performance genre, and shopping as a leisure activity.

The intensive uses of recorded sound are many. Some of the most intensive have been and remain highly idiosyncratic, practiced only by certain subcultures or subgroups of users. The most avid collectors, for instance, obsessively search out and possess records in addition to listening to them. So-called audiophiles obsessively attend the sound quality of playback. And DJ's scratch, sample, and mix records. The avid fans and fan clubs associated with particular recording artists and musical styles can be just as intense in different ways; indeed, the term *fan* derives from *fanatic.* These uses and users are idiosyncratic in the sense that none of them merely play recorded sound, if the word *merely* can be used to denote expectations of extensive use that attend mass media. But recorded sound was and is intensively consumed in less idiosyncratic ways too, by everyday users who *do* "merely" play records. In particular, repetition represents another significant form of intensity. Though certain users (small children and professors, according to Roland Barthes) have a greater investment in repetition than others, part of the habitual intensity of using recorded sound is repeated play.[15] Part of the practice of "merely" playing records is playing them again and again.

Just as extensive, industrial print production and "speaking" on paper provide important contexts for understanding the first, primitive and ephemeral phonograph records, the history of another, more specific print commodity can offer one helpful context for understanding the emergence of recorded sound as a unique mass medium. Nickel-in-the-slot and exhibition phonographs enjoyed great popularity for several years in the early and mid-1890s without becoming a genuinely mass phenomenon. A best-selling record meant sales of something like five thousand copies over two years; the first "million-selling" records are usually ascribed to 1903–1905.[16] By contrast, print media already enjoyed a mass audience of long standing. As early as the 1820s (for the American Bible and Tract Societies) and the 1830s (for the urban penny press) reading audiences in the United States ranked toward the millions rather than the thousands. Beyond mere numbers, however, there were qualitative changes to print media in the mid-1890s that signaled a new kind of market. In *Selling Culture,* Richard Ohmann (1996) argues that U.S. mass culture arrived first in the pages of magazines like *Munsey's, McClures,* and *Cosmopolitan.* He sees mass circulation monthly magazines like these as the cardinal form of mass culture in the United States because, starting around 1893, a growing number of such publications integrated additional illustrated advertisements into their feature pages, and started to profit much more on the sale of advertising revenue than on the sale of issues and subscriptions. This decisive reorientation—from selling magazines to seducing consumers and "selling their eyeballs"—had no exact counterpart in the history of recorded sound, at least until the commercialization of radio in the 1920s, but it nonetheless helps to contextualize the new medium.

The timing, scale, and scope of the modern monthlies all make them helpful yardsticks. Simply in terms of numbers, the aggregate circulation of monthly magazines shot from eighteen million in 1890 to sixty-four million in 1905. In terms of content, scholars generally agree that the magazines helped map the social spaces of U.S. life in which "women were usually singled out as the trainees for participation in the commodity-laden modern world" (Bogardus 1998, 518). Advertisers pitched to women in the women's magazines and the general circulation ones, as the vague category of "customer" itself became increasingly gender typed.[17] Indeed, the National Phonograph Company advertised in *Munsey's* as early as 1900, while the Victor Talking Machine Company had begun its lavish advertising campaigns in *Cosmopolitan* and the *Saturday Evening Post* by 1902. In 1906, Victor boasted that its "advertising campaigns reached some 49 million people every month," more than half the U.S. population, while Edison's reputedly less aggressive National Phonograph Company advertised its wares by placing full-page ads in more than a dozen national circulation magazines each month, including *Cosmopolitan, Munsey's, Good Housekeeping, Everybody's,* and *Outlook.*[18]

More than simply a platform for advertising home phonographs, the modern monthlies helped enable and were enabled by some of the very social, economic, and cultural conditions that helped make home phonographs such a success. If the "big three" phonograph companies—Victor, National, and Columbia—started their meteoric rise roughly three years after the new *Munsey's, McClures,* and *Cosmopolitan,* they nonetheless joined the modern monthlies as, in Ohmann's (1996, 29) terms, a "major form of repeated cultural experience for the people of the United States." By 1909, the phonograph industry was producing 27.2 million records a year, still a fraction of the aggregate circulation of the magazines.[19] Yet while monthly issues had a shelf life of one month, phonograph records individually survived on a logic of repetition. Even more than print media of the time, records were *repeated* cultural experiences, literally played again and again and again. This distinction seems central to the meanings of the home phonograph as an element of mass culture and is related to nonprint commodities as well. When a woman took down a box of Uneeda biscuits, the brand name was familiar, and the biscuits were continuous with the contents in previous tins or packages. All Uneeda biscuits looked and tasted the same; all fifty-seven varieties of Heinz pickles were consistently, coherently different from one another. That sameness and coherence formed part of the magic of standardized mass production. It was "magic" partly because as much as the biscuits and pickles of each variety might look and taste the same, they really were different.[20] By contrast, phonographs and phonograph records helped to introduce the intensity of true repetition to the performance of mass markets. Leaving aside the question of whether two records of the same sounds were really the same, individual users of recorded sound continually reused their own same records without ever—or hardly ever—using them up.

Even F. W. Gaisberg (1942, 18), who performed on early records for Columbia and then traveled as (what would later be called) a talent scout and recording engineer, found himself compulsively repeating the same record over and over: "The thirst for music among the people must have been prodigious to endure the crude and noisy records produced at that time. I remember my own affection for those rough tunes. I seemed never to tire of repeating the record of 'Ben Bolt' from *Trilby.*" Gaisberg's wonderment betrays his assumption that "the people" shared a thirst for recorded sound that was both impossible to slake and at the same time irrational, focused inappropriately and obsessively on poor-quality recordings because of the melodies they contained. Whereas rational consumers might grudgingly "endure" the noise, the contemporary marketplace induced even those who—like Gaisberg—should know better into an enervated cycle of repetition: "I seemed never to tire." Gaisberg's repetition of "Ben Bolt" was but one instance of "Trilbymania," the remarkable fad that gripped the United States after George Du Maurier's popular novel *Trilby* was serialized by *Harper's* in 1894. (In the novel, the character Trilby sings "Ben Bolt" before she is hypnotized by the evil Jew, Svengali, when she can also sing a more sophisticated repertoire.) While Gaisberg played and replayed his record, there were twenty-four theatrical versions of *Trilby* playing on the U.S. stage, including parodies, and consumers could also choose between Trilby hats, Trilby dolls, Trilby shoes, and other Trilby goods that vied for their attention. The *New York Times* lamented that everyone had "to know 'Trilby,' to talk 'Trilby,' to eat 'Trilby,' [and] to dream 'Trilby.'"[21]

This so-called Trilbymania has been puzzled over by cultural historians. Like fictional Trilby herself, consumers were transfixed. As Susan A. Glenn (2000, 91) explains, the craze is "testimony to the cultural centrality of questions of selfhood" at the time, "not only the vulnerability and mutability of the female self, but also the larger question that faced both sexes confronting an emerging industrial society where hypnotic suggestibility might be induced by the lure of material goods, by the manipulations of advertising, or, as some social theorists worried, by the psychological sway of the crowd or the mob." *Trilby* and Trilbymania together offer a parable of mass consumption, complete with a (now familiar) raucous intertextuality—a swirl of meanings that connect across magazines, books, sheet music, drama, and dry goods—and inflected by an absentminded yet virulent anti-Semitism, since the Dreyfus affair was playing out in the press at exactly the same time and shared the partly Parisian setting of Du Maurier's novel. While Americans read, listened, and shopped, the popular "Ben Bolt" was variously "sung by" fictional characters, stage actresses, recordings, and (doubtlessly) consumers themselves. Gaisberg may or may not have sung along, but wearing down his record of "Ben Bolt," like consuming other Trilby goods, produced its own meanings according in part to the mode, frequency, *and reproducibility* of the experience. Gaisberg's "Ben Bolt" repeatedly made sense to him in the contexts of its own repetition as well as in the combined contexts of Gaisberg's life, his self, and the ambient mania.

The sort of repetitive intensity that Gaisberg and other phonograph users indulged in had previously been more a feature of musical education ("Practice, practice, practice") than musical reception. It was reminiscent of the literacy practices surrounding devotional texts, for instance, or literacy in situations of particular scarcity, when a single newspaper or mail-order catalog got read intensively, again and again, and by many readers. Consumers today have gotten used to the way in which small children play the same videocassettes over and over (and over and over and over) again, or the way some idiosyncratic cultural forms seem to elicit idiosyncratic repetitions (*It's a Wonderful Life* and *The Wizard of Oz,* for example), but adult U.S. culture more typically consumes and discards, reads and recycles, buys extensively and then buys some more. Phonograph records, then tapes, videocassettes, CDs, DVDs, and MP3s also counter that trend; part of the complex logic they harbor as material possessions is repetition and continual reconsumption, rewind and replay.

I will return briefly to this question of repetition and the role that almost ritualized repetition seems to have played in the social construction of the home phonograph, amid the magic and desires of the modern marketplace. First, however, it is necessary to think more directly about the domestication of mechanical reproduction. If users were intensively reusing or repeating records, the records they played were also elaborately constructed as repeat performances, as the mechanized reproduction of desired sounds. The language of *mechanical reproduction* is not anachronistic in this case. It is possible to call phonographs a form of reproductive technology with assurance because one crucial part of every phonograph was its "reproducer," incidentally containing a "diaphragm"—the parts that resonated in response to grooves on the record and thereby reproduced sound. The term *reproducer* of necessity entered the vocabularies of many phonograph owners at the turn of the twentieth century. And if the new medium of recorded sound thus provoked little changes or additions to the semantic lives of Americans, it likewise came to possess meanings within and against existing discourse more broadly defined. Sound reproduction became defined by and against an existing field of metaphors, attitudes, assumptions, and practices. Varied constructions of mimesis and music formed important contexts for the uses and users of the new medium.

The vocabulary with which phonographs were introduced and the symbolic terrain they occupied were all part of the definition of the medium, its coming into focus, first as a novelty and eventually as a familiar within U.S. homes, embodied in a phonograph that sat right near where the radio and then the television would ultimately sit further on into the century. Like the discursive lives of those later media, the discourses making sense of recorded sound formed a matrix of heterogeneous, changing, and even contradictory messages. These messages were registered in part within promotional representations—advertising, trade brochures, published accounts, and the habits of retail establishments handling the products. Also like radios and televisions, part of the discursive life of the phonograph emanated from the design of the machine itself and its location within the home.[22] The japanned surfaces of an early tabletop machine or the mahogany finish of an enclosed-horn Victrola (1906) were each suggestions of the way a machine might fit into home decor. Additional messages lay coded into records themselves, which reproduced music yet also offered effective *representations* of music, whether implicitly or explicitly. Records rendered two- to three-minute versions of—metonyms for—a genre, composition, and performance, packaged materially and acoustically for private consumption.

"Band" records were actually recorded by small ensembles representing bands; recorded musical pieces were short segments or pastiches representing whole compositions; comic sketches were two-minute records representing whole fifteen-minute vaudeville "turns"; and the earliest recordings were announced and even occasionally applauded, representing live performances.

Figurative representations of phonographs and records underwent a particularly important change as part of the redefinition of recorded sound as a form of domestic amusement. The metaphors of inscription and personification that had initially helped to define the medium were gradually displaced, and then replaced by richer metaphoric identifications of the phonograph within the existing discourses surrounding music and home in U.S. life. When Edison first unveiled his phonograph at the New York offices of *Scientific American,* he and witnesses alike anthropomorphized the device, which saluted them and asked after their health. A decade later, a program distributed at Worth's Palace Museum in New York urged novelty seekers, "Before leaving the museum don't fail to interview the wonderful EDISON PHONOGRAPH." Americans stood ready to personify new technology. Yet somehow these metaphors did not follow the phonograph into U.S. homes. Playback did not elicit the same personifications that recording-plus-playback did. Although the earliest phonographs as well as those promoted for office use were routinely understood and represented according to metaphors of writing and embodiment, the home phonograph was not. Writing metaphors continued to matter in the specialized context of intellectual property disputes over recording (since the U.S. Constitution explicitly protects "writings"), while personifications lingered only in the commercial literature surrounding dictation phonographs. The dictating machine was first promoted as a businessman's "ideal amanuensis," gendered male. A few years later, when women made up more of the nation's office workforce, the cover of one National Phonograph pamphlet made a simple equation by picturing a phonograph beside the words "Your Stenographer," while corporate propaganda assured wives that their businessmen husbands were dictating to a phonograph, "instead of talking to a giddy and unreliable young lady stenographer." Home phonographs simply did not elicit the same personifications. That makes them unusual. Cars and boats remain "she," while many early domestic appliances, including home electrification, were sometimes represented in terms of domestic servants or even slaves.[23]

Part of what helped to occlude or deflect metaphors of personification in the representation of home phonographs was the continued instability of phonographs and records as mimetic goods. The problem of representing them as representational proved vexing, judging at least from the catalogs and advertisements for amusement phonographs and related supplies that make competing as well as confusing claims of verisimilitude. A strangely varied and inexact language of exactness came to dominate promotional representations of the medium as a means of representing music. Victor advertisements soon assured readers that "the living voices of the worlds' greatest artists can now be heard, whenever you choose, in your own home." Edison records were by contrast "the acme of realism." This Victor copy suggests complete transparency, access to the actual and natural voices of performing artists. But the Edison slogan boasts of supreme artifice, the ultimate reality effect. And similar confusions arose and remained throughout the literature.[24]

With the displacement of metaphors of personification by varied mimetic claims made on behalf of phonographs and phonograph records, gender difference became newly and more literally instrumental to representations of sound recordings itself as an adequate and appealing means of representation. Women's voices formed a kind of standard for recording because they proved particularly hard to record. Despite its early success with band music, Columbia was utterly unable to record women's voices well as late as 1895, when Lilla Coleman's records were admitted in its catalog to be "suitable only for use with the tubes—Not adapted for horn reproduction." The Boswell Company of Chicago offered its "high grade original" records in 1898 with the assurance that "at last we have succeeded in making a true Record of a Lady's voice. No squeak, no blast; but natural, clear, and human." The Bettini Phonograph Laboratory in New York similarly claimed "the only diaphragms that successfully record and reproduce female voices." The Victor Company offered a few sopranos, but contralto voices were the norm for the few women recorded on the earliest records. As late as 1904, one phonograph handbook warned that "taking high notes" could be troublesome, particularly in the case of "ladies' voices," which tended to render "a harsh screech, technically called [a] 'blast.'" Part of the tacit knowledge accrued by recording engineers like Gaisberg was the ability to record different voices, timbres, and instruments successfully (sometimes through "subterfuge"), despite the limitations of acoustic recording technology.[25]

Film theorist Richard Dyer (1997, chapter 3) has explained the way that film lighting historically normalized white skins, making the filmic reproduction of nonwhite complexions the special or "abnormal" case. Recorded sound, somewhat like telephony, provides a related (if inverted) instance at the turn of the twentieth century: mediated sound was normalized in relation to women's voices. As one German official rationalized in 1898, telephone companies had to rely on women operators in part because the frequency of their vocal range was "especially well suited" for transmission, making them "more easily understood than men."[26] Similarly, it is clear that early recording technology succeeded *as* recording technology according to its success with women's voices, particularly those of sopranos. Recording women's voices so they sounded "natural, clear, and human" proved that recording worked, as the nature of records and women coincided. The visual and aural mimetic codes attending modern media, in other words, are constructed partly of racial and gender differences—differences that habitually attend users, not publics. Nonwhite skins and women's voices became particularly potent indexes of "real" or successful representation, though of course success (like realism) varies according to the social and perceptual conditions of each medium as well as contemporary aesthetic norms. Filmic representations succeeded amid cultural constructions of race as a form or source of visual information. Telephonic representations succeeded amid constructions of "the operator" as both gendered and effaced, available to facilitate transmission but hardly to transmit. And phonographic representations succeeded amid constructions of soprano voices as desirable commodities in themselves. Not that such visual and aural mimetic codes were the exclusive, static, or entirely thoughtful conditions of respective media: the Bettini Phonograph Laboratory adopted as its slogan "A True Mirror of Sound," mixing its metaphors as if aural mimetic codes might be effectively constructed or translated from visual ones. (Similarly, one maker of piano rolls in the 1920s charmingly advertised its reproducing rolls as "the film of the music camera.")[27]

Just as Boswell Company records were reputedly "original," Bettini's were "autograph records," the telling inscriptions of unique human voices. Both terms meant to indicate that these records were recorded from human voices rather than duplicated from preexistent recordings, which was probably a common practice in the 1890s. The distinction between original and nonoriginal records may have confused consumers, who were necessarily more mindful of the broader distinction between live music and recorded sound. Slippage in terms like original, true, natural, living, and real in the literature promoting home phonographs served to emphasize rather than contradict the apparent power of mechanical reproduction to appeal and enthrall: everywhere Victor's trademark dog, Nipper (trademark 1900), sat listening for "his master's voice." The pleasures of that slippage, the contiguity and contestation of imitation and reality, are evident in the mass circulation of Nipper's image as well as in the records themselves. Many of the earliest records were marketed without identifying the recording artists who performed them. Some of Columbia's earliest artists made recordings that sold under many different names, as if made by many other people. Bettini, which did identify well-known bel canto singers of the day, also offered records of a "Lady X," coyly represented in its 1898 and 1899 catalogs

26

LADY X

POPULAR AMERICAN SONGS AND NEGRO MELODIES

Price, $1.25 each

1	I Want You, Ma Honey	*Fay Templeton*
2	All Coons Look Alike to Me	*E. Hogan*
3	Louisiana Lou	*Leslie Stuart*
4	Yvette	*J. L. Golden*
5	The Harmless Little Girl (From Lady Slavey)	*Kerker*
6	A Hot Time in the Old Town	*Theo. A. Metz*
7	The New Bully	*Trevathan*
8	My Coal Black Lady	*W. T. Jefferson*
9	I want Them Presents Back	*Paul West*
10	Little Alabama Coon	*Hattie Star*
11	Mr. Johnson, Turn Me Loose	*Ben. R. Harney*
12	My Gal is a High Born Lady	*Barney Fagan*
13	Mamie, Come Kiss Your Honey Boy	*May Irving*
14	Isabelle, A Girl Who is One of the Boys	*Bratton*
15	A Simple Little String (From the Circus Girl)	*L. Monckton*
16	I Don't Love Nobody	*Lew Sully*
17	Frog Song	*G. E. Trevathan*
18	Standing on the Corner	*Geo. Evans*
19	My Baby is a Bonton Belle	*C. L. Davis*
20	The First Wench Done Turned White	*Ed. Rogers*
21	Lou, Lou, How I Love My Lou (from Pousse Cafe)	*Mills*
22	Baby (from the Lady Slavey)	*G. Kerker*
23	He Certainly Was Good to Me	*A. B. Sloane*

Figure 2.1 "Lady X" performs for the Bettini Phonograph Company, 1898 and 1899. (*Source:* Library of Congress.)

with her back turned to conceal her identity; she recorded "Popular American Songs and Negro Melodies." Because recordings displaced the visual norms of performance (listeners couldn't see the stage), they hinted at imitation or ventriloquism in new ways, just as mimicry was enjoying such great popularity in U.S. vaudeville and musical theater.

Mimicry was the particular province of vaudeville comediennes like Cissie Loftus, Elsie Janis, Gertrude Hoffmann, and Juliet Delf. One reviewer even delighted in a contest between the "Cissie Loftus Talking Machine" and the "Gertrude Hoffmann-ograph," when the two headlined on the same bill, competing to see which of them could imitate each other's imitations best. Personification was stood on its head, as women were facetiously celebrated as machines. Such mimicry and its enthusiastic reception helped open "questions about the relationship between self and other, individually and reproducibility" that proved both provocative and timely.[28] As Susan Glenn, Miles Orvell, and others have described, U.S. culture at the turn of the twentieth-century was deeply engaged with questions of authenticity and artifice, realism and illusion. There were celebrations of certain imitations as potently "true," while in literature and the other arts, "the real thing" proved an elusive category, pleasurably attended in its elusiveness. "Cissie Loftus Is Not a Phonograph," another reviewer had remarked in 1899, while Loftus herself sometimes created her imitations with the help of recordings, and in one of her vaudeville turns even played a record of opera star Caruso before imitating him by imitating it.[29]

Offstage mimicry thrived as well. Manufacturers urged consumers to "accept no imitations," although such warnings routinely went unheeded. In the music trades, the three big phonograph companies were harried by pirates, as competitors sprang up and tried to "dupe" (duplicate) records, re-create successful recordings, or undercut prices. At the same time, more than half of the pianos sold in the United States were reportedly "stencil" instruments, labeled and sold by companies that had not manufactured them (the particular bugaboo of Steinway, Chickering, Baldwin, and other famous makers). Of course, the preeminent claim of verisimilitude available to phonograph promoters and listeners alike was the surprisingly pliable notion of acoustic fidelity. Recordings sounded exactly like the sounds they recorded, although the quality of sounding "exactly like" has continued to change over time and according in part to available technology, most recently from the standards of analog to those of digital recording. Any mechanical form of sounding "exactly like" must have been defined in part against popular vaudevillian mimicry as well as in light of the amateur mimics who probably emulated (that is, mimicked) well-known vaudevillian mimics at home or in their own social circles. "Singing along" with recordings came to make unnoticed sense as "singing like" recording artists.

In addition to tapping the varied discourses of U.S. realism and mimesis, home phonographs gradually came to make sense against and eventually within the musical practices of the day. To give anything like a complete summary would be impossible, but there are certain "givens" regarding U.S. musical life at the turn of the twentieth-century, among them the association of home, woman, and piano, and the complementary though possibly less portentous association of outdoor public space, man, and band music. Both were to be tested by the immense popularity of recorded band music for home play.

Music literacy rates were high. Among the middle and upper classes some level of musical literacy was expected of all women, and those talents were freighted with the sanctity of home and family. Hundreds of companies made pianos to feed these expectations; the industry managed to sell 170,000 pianos in 1899 alone, and numbers kept climbing. Meanwhile, there were tens of thousands of band members around 1900, many professionals but most amateurs, their gathering, practicing, and playing the evidence of communities fostered by civic, ethnic, or institutional identities. Music of all kinds had recognized social functions, gendered relations, and moral valences. Opera, in particular, was both the subject and instrument of (high/low) cultural hierarchy. Pianos were both the subject and instrument of (middle-)class aspiration. Ragtime was both the subject and instrument of quickening markets and (racialized) play. And mimicry was both the subject and instrument of (feminized) explorations of selfhood.[30]

Clearly, the arbiter of musical activity within the home was the woman, while the most direct arbiter of musical activity at large was most likely an uncalculated combination of sheet music publishing houses, metropolitan performance institutions, and an army of roughly eighty thousand music teachers of both sexes (81 percent women by 1910). Increasing professionalization was applauded on civic and national levels, while the professionalization of women was usually condemned. (Of roughly fifty-five thousand professional musicians in 1910, 71 percent were male.) Musical periodicals carried chastening stories of popular divas and their harrowing lives, while mass-circulation monthlies like *Good Housekeeping* lamented when any young woman, suffering from too much talent or ambition, returned from a conservatory and denied "to her father and mother the simple music that they love and understand." ("She has learned that Beethoven and Chopin and Schumann are great, but she has not realized that simpler music has not lost its charm. . . . Perhaps she has caught Wagneritis.") To some observers, women were simply condemned to amateurism. James Huneker, a writer fond of sorting European composers into masculine and feminine types (Bach and Beethoven versus Haydn, Chopin, and Mendelssohn), summed it up, "*Enfín:* the lesson of the years seems to be true that women may play anything written for the piano, and play it well, but not remarkably."[31]

It helped not at all that the most successful popularizer of "good" music in the era, bandleader John Philip Sousa, was both prone to a noticeably "feminine" fastidiousness and explained his popular repertoire as an act of redeeming the fallen. Played by Sousa and his men, a "common street melody" became a respectable woman: "I have washed its face, put a clean dress on it, put a frill around its neck, pretty stockings, you can see the turn of the ankle of the street girl. It is now an attractive thing, entirely different from the frowzly-headed thing of the gutter."[32] Thus Sousa popularized good music and made popular music good. In his several perorations on the "menace of mechanical music," Sousa deployed similar metaphors to equal effect. The pianola and the phonograph, he was sure, would reduce music to "a mathematical system of megaphones, wheels, cogs, disks, cylinders . . . which are as like real art as the marble statue of Eve is like her beautiful, living, breathing daughters." To use these devices was to subvert nature in a world where naturalness and womanliness coincided with seeming ease; "The nightingale's song is delightful because the nightingale herself gives it forth." Sousa warned that these machines were like the recent "crazes" for roller skates and bicycles, but that they might do more damage, like the English sparrow, which "introduced and welcomed in all innocence, lost no time in multiplying itself to the dignity of a pest, to the destruction of numberless native song birds." Sousa's metaphors drift confusedly amid gender and national categories in their allusion to birds and description of musical culture. Women amateurs have "made much headway" in music, he wrote approvingly, but the mechanical music will make them lose interest, and "then what of the national throat? Will it not weaken?" Sousa's U.S. amateur loses some of her gender definition in his next question: "What of the national chest? Will it not shrink?" Whatever else he imagined, Sousa foresaw the coming decline of amateur music-making with great perspicacity.

In all of its modalities—performance, instrumentation, composition, and education—the sounds, subjects, and spaces of U.S. music were shot through with assumptions of moral and aesthetic value that remained inseparable from active categories like tradition, class, race, gender, domesticity, and professionalism. What interests me here are the translations that became available between and among categories around 1900, which might indicate points of contestation or change in the mutual discourses of music and home where the new medium could take root. Among them there were public, performative translations, of course, like Sousa's play across the categories of popular and "good" music, or like the adaptive traditions of blackface, which played across and against categories of race, class, and gender. But there were other translations across other categories as well, and the home phonograph became party to many. Particularly evident, for example, was slippage in the relative operation of the categories amateur and professional against the categories domestic and public. Victor advertisements asked, "Why don't you get a *Victor* and have theatre and opera in your own home? The *Victor* is easy to play" (1902), while National Phonograph assured that its product "calls for no musical training on the part of any one, yet gives all that the combined training of the country's greatest artists give" (1906). Both appeals resemble contemporary advertisements for pianolas and player pianos, which stressed the ease of play along with the salutary musical production—good for the soul, good for the family.[33] At work was a partial translation between amateurism and professionalism that tended to enforce the amateurism of home listeners, not just in the subsequent withering away of live home music, as Sousa so astutely recognized, but also in the celebrated availability of professionally produced music in the home. Records and piano rolls were "professional" in the dual sense that they reproduced the work of paid musicians and were the standardized, mass products of purposeful corporate concerns with which listeners were now engaged in commercial relations.

Opera records were particularly suggestive of available and varied translations among high and low tastes, professional and amateur music, and live and recorded performance. Two-minute opera records had about as much to do with actual operas as the Vitagraph Company's fifteen-minute Shakespeare films had to do with actual plays by Shakespeare in the same period.[34] "Opera" and "Shakespeare" were porous categories active in contemporary constructions of class and public life. At a time when any small town might boast an opera house, "opera" signified much more than a single musical form. *Allusions* to opera operated as cultural currency, circulating as "consensus builders" and "distinction makers" in different contexts.[35] By a similar token, as William Kenney (1999, 45) explains, opera records often included snippets of arias, but they also frequently "presented operatic voices interpreting traditional and folk songs." Even popular songs could be offered on opera records, if performed by musicians trained at a conservatory. The result was an even more porous taste category, marked anew by distinctions between amateur and professional music for "high-class" listeners at home. Opera was newly a style of material possession, rather than just a kind of music, and opera records apparently did much to elevate the medium of recorded sound in the public eye, even if popular music and other "Coney Island" fare formed the real bread and butter of the incipient recording

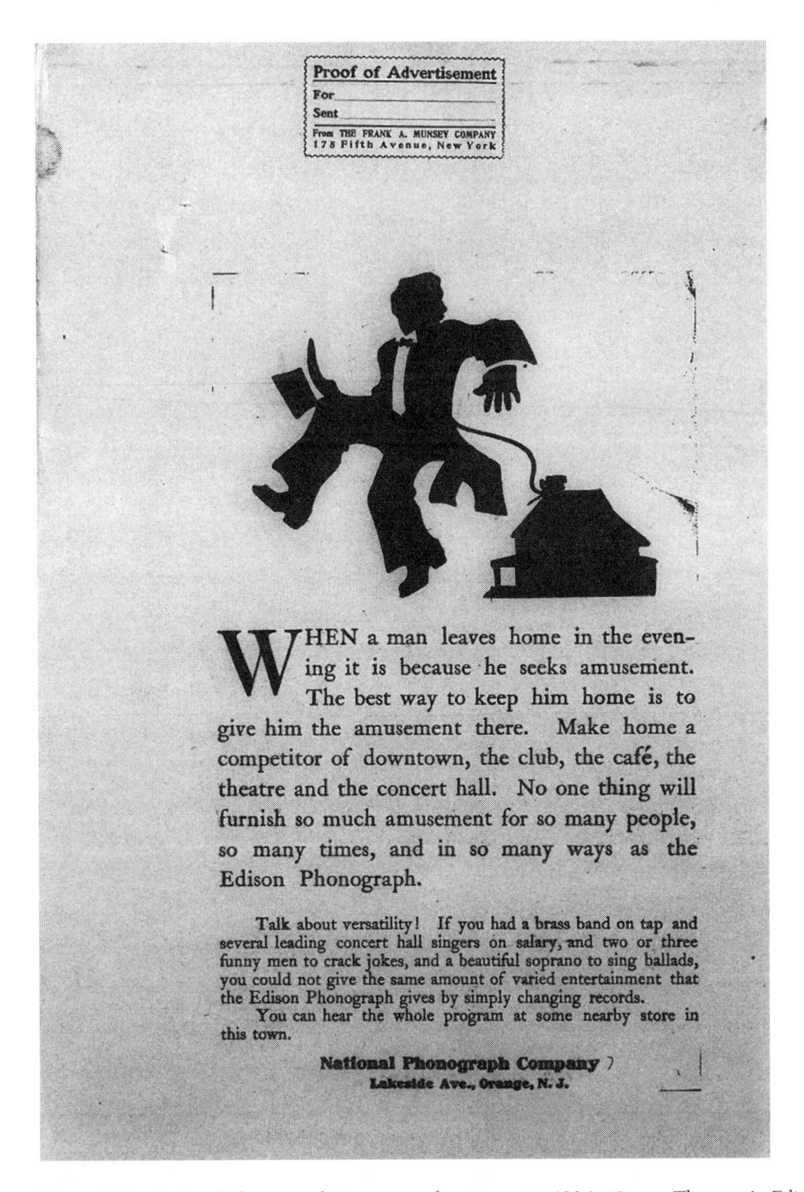

Figure 2.2 National Phonograph Company advertisement, 1906. (*Source:* Thomas A. Edison Papers Microfilm Edition.)

industry. The Victor Company, in particular, succeeded in promoting itself "for the classes," leaving Edison's National to admit that it served better "for the masses."[36]

The uses of recorded sound thus involved the use and adaptation of a host of different categories and oppositions. If opera records offered a way to experience the real, the professional, and the high class, they also offered the experience of those categories in variation, flux, and dynamic cross-indication. Nowhere is this phenomenon more evident or striking than in the repertoire of "ethnic" or "old-time" recordings offered by all of the major record companies.[37] Phonograph records were the first widely available, mass-produced goods categorized according to ethnicity and national origin. Mail-order catalogs and the new department stores already divided everything from underwear to pocketknives into age and gender categories, the Heinz pickle company boasted those fifty-seven varieties, while premium imported goods had long been associated with points of origin: Egyptian cotton, Turkish tobacco, French perfume, and Ceylon tea. But the National Phonograph Company, Columbia, and the Victor Talking Machine Company alone offered thousands of products according to national and ethnic type. In one sense, the pocketknives and phonograph records offered a map of U.S. differences, produced by manufacturers and retailers, read and reproduced by consumers. But in another sense, no map of differences differing—of categories in variation, flux, and dynamic cross-indication—could ever be drawn.[38] Again, *using* records is what mattered most, though in the case of ethnic records, many users potentially comprised "counterpublics," in Michael Warner's (2002) terms, instead of disaggregate and nonideological "users," in the sense that this chapter elsewhere promotes. That is, the users of ethnic records "belonged" accordingly to one ethnicity, to one niche market or the next, although such belonging remained frayed at several levels by the problems and pleasures of category variation, flux, crossovers, and cross-indication.

In the repetitive formulas of their monthly and annual lists, record company publications represented cultural identities, both by dint of the many and elaborate distinctions drawn among those identities and by dint of promoting as representative the musical selections pertaining to each group. Selections could be representative of some purportedly authentic national or ethnic essence, or expressive of a parodic genre or burlesque (typically relying on dialect humor) widely recognized as evocative of us/them distinction drawing itself. Glimpsed within these early record catalogs, music was divided into types, many of them mutually exclusive and ethnically or nationally based. There was Hungarian music, Bohemian music, Hawaiian music, and Hebrew music, as there must correspondingly be Hungarians, Bohemians, Hawaiians, and Jews. There was Italian music and other music in Italian, somehow intricately leveraged by the existence of both Italians and opera aficionados. There were plantation melodies, so-called Coon songs, and Negro lyrics, operating within a cultural matrix at least as complex as the one associating Neapolitan airs and Caruso arias.

The cultural data of phonograph records was importantly a *matter* of representation. It was equally significant that records purported to represent authentic cultural identities, that they belonged to one category or another, and that they were material goods or belongings, matter, available to the hands of individual consumers. In many respects it was their physical quality as standardized, mass-produced goods that helped to enforce their quality as specific cultural data, even as the culture they represented proved variable and unspecific in the extreme. Different music, records, and sampled identities were differently charged with cultural difference. There were eventually Hungarian records with lyrics in Magyar as well as many versions of Franz Liszt's widely recorded "Hungarian Rhapsodies." Even Richard K. Spottswood (1990, 1:xvii), the eminent discographer of ethnic records in the United States, has found it impossible to pull all of the relevant categories apart from one another.[39] While early on the Victor Company listed a few dozen Chinese titles in Mandarin and Cantonese for Chinese consumers, the wholesale merchandising of Hawaiian music in the same period had much more to do with non-Hawaiian consumption of "Hawaii" and "Hawaiian" culture. Immigrant niche markets coexisted with catholic tastes and crossover hits.

What I am suggesting is that phonograph records frequently proved transgressive of the very cultural categories that they helped to represent as distinct or specific. Using the new medium offered *inter*cultural experiences of varying intensity in addition to cultural experiences of varying weight. While this is perhaps more of a comment on the social functions of popular music than it is on phonograph records, playing records must have engaged users within a range of experience relevant to cultural difference. Playing records is notoriously difficult to document, of course, but research by Lizbeth Cohen (1990, 105), Victor Greene (1992), and others shows that consumers of "ethnic" records and like commodities used them to negotiate their inclusion within a wider public—a national audience or U.S. identity—and at the same time to maintain or activate potential distinctions between that public and the immigrant counterpublics to which they belonged. What I want to also emphasize here is the ways that material goods work as potent actors within and among dynamic social groups and individuals, able to mark categories of difference and engage issues of categorization as such, whether those categories are ones like "Hungarian," "high class," and "Hawaiian," or ones like "home."

"There's no place like home." "Home sweet home." The sentiment, like the clichés, seems eternal. It should come as no surprise, though, that against the welter of cultural change alluded to so cursorily above, the cultural construction of U.S. homes was itself undergoing a process of dramatic change that is relevant to the domestication of the new medium. The Edison trade name "Home" phonograph, the National Phonograph Company's most basic model, rings with changing and complementary conceptions of home and phonograph. The middle- and upper-class parlor with its piano was becoming a "living room," as U.S. homes became more expressive of the personalities of their inhabitants. Accordingly, public space was evolving as well, partly as the result of an increasingly urban, increasingly mobile population and a growing number of women in the workforce. Changes to public space became evident in things like park construction, the rapid spread of public amusements, a growing tissue of outdoor advertising, and changes to the patterns of retail and habits of outdoor recreation.[40]

Consider, for one, the social spaces where people shopped. The Victor Talking Machine Company erected a huge electric sign above Broadway at Thirty-seventh Street in New York City in 1906. Visible from Madison Square three-quarters of a mile away and illuminated at night by one thousand lightbulbs, the sign read "Victor" above the usual picture of Nipper. Below the caption, mistakenly made plural in this instance—"His Masters' Voice"—the sign continued in seven-foot letters, "The Opera At Home." The company boasted that eight hundred thousand men and women saw the sign every day. (Many of the same number must have seen it repeatedly, the next day and the next.) It loomed two blocks north of the new Macy's at Thirty-fourth Street and two blocks from the old Metropolitan Opera House on Seventh Avenue at Thirty-sixth Streeth. The sign and its location are suggestive. The "Opera" advertised in gigantic letters "At Home" could not but evoke and resemble the more sedate "Opera" between "Metropolitan" and "House" a few steps away. Stars at the Metropolitan were already cutting records, to be sure, yet there was no simple conversion of Opera House into Home Opera, in large part because the terms of such a conversion were contested by the public and commercial nature of its suggestion. The big electric word "Opera" seen by eight hundred thousand moving people already violated a central precept of opera as a taste category or a performance of status definition for a comparatively select few. This "Opera" had as much to do with Macy's, which aggressively sold Victor goods, as it did with the Metropolitan. And it had plenty to do with popular music, which remained a staple at all of the record companies, despite commercial paeans to opera and classical music. Likewise, the gigantic "Home" could not signify a family abode, a refuge from urban chaos, without calling on the public spaces

Figure 2.3 "His Masters' Voice" in Herald Square, 1906. (*Source:* New York Public Library.)

that served to inscribe, if not jeopardize, that sanctum—among them the workplace, street, and store. Then the image of Nipper, as difficult to parse as it was apparently compelling, loomed all the more confusing in the plurality of his "Masters'" unitary "Voice." Was Nipper at "Home"? Who were his "Masters" there? And how was their one "Voice" reproduced on the record player that sat beside him? These unasked and unanswerable questions at once recall the slippage in descriptive terms like "real" and "live" as they were applied to recorded sound, and demonstrate the extent to which the translation from public to private remained shot through with power relations, indeterminate evocations of taste hierarchies, social superiority, mastery, and seduction, all tied intricately to the immense power of the interplay between mimesis and mechanical reproduction.[41]

The same translations were necessarily evoked inside stores like Macy's, where the "dream worlds" of mass consumption beckoned.[42] Department stores were not the only businesses to sell phonographs and records, however. Phonographs and records were sold in music stores, from the gigantic Lyon and Healy firm in Chicago to small-town shops specializing in sheet music, lessons, and instrument repair. And they were also sold in stores where hardware, sporting goods, or dry goods were the main articles of trade. In each of these venues, phonographs and records helped to further theatricalize the point of sale. Without radio to familiarize listeners with new songs and recordings, phonograph demonstrations were a necessary part of every shopper's curiosity and desire. "Pluggers" (and payola) tried to influence sheet music sales in music stores and at the music counters of the big department stores. Demonstrations were a recent, if familiar, part of selling everything from Fuller brushes to cosmetics. Phonographs and records put the two together, helping to ensure that home play was replay, the repetition of a public and commercial desire along with its translation into related, private, personal reenactments. Lyon and Healy offered "concerts" every day, free and open to the public; a live pianist performed, but most of the music came from a Victrola, playing to tired shoppers and lunchtime idlers in Chicago's Loop. Smaller stores sometimes organized "recitals," but they were almost always prepared to play sample records on request as well.[43]

Everyday retail practices and contexts can be hard to document, but there is some revealing data for the National Phonograph Company. Faced with a legal challenge to its sales rights in New York State, Edison's company did a survey of its upstate dealers in 1906. That was a boom year for cylinder phonographs, and the survey offers a rare look at local sales operations. Out of 133 dealers visited (some of them also wholesale jobbers), it was notable when one, like William Harrison in Utica, devoted his or her business to phonographs and records exclusively. In Watertown (pop. 27,787), there were 7 dealers, one specializing in "stoves and household goods," and another in "wallpaper, mouldings, etc." Many music stores carried phonographs, though some were notably discouraged "that it affects the piano and musical end of their business." In Buffalo, there was a drugstore selling phonographs out of a back room; in Elmira, the Elmira Arms Company was doing well; and in Syracuse, a furniture store was struggling. In Oneonta, one tiny dealer "keeps Edison phonographs and records to accommodate his customers who are mostly farmers"; "he says when they come to his place for records they are liable to purchase other goods that they might require." Most carried small stocks of machines and records, and all save the one dealer in Cobelskill (pop. 2,800) had competition from other Edison dealers in the same town, plus the dealers pushing Columbia and Victor goods.[44]

Some handled competing brands. One common situation was a bicycle or sporting goods store that specialized in phonographs during the winter. There was the Utica Cycle Company, the Rome Cycle Company, and George W. Johnson of Rochester, who "May first of each year takes his phonographs from the windows and puts in bicycles and on October first each year he takes his bicycles from the window and puts in phonographs and records." Even the boosters at the trade journal *Talking Machine World* (1905) soon acknowledged the strategy. The magazine initiated a Sidelines column in 1908, and suggested that dealers branch out into illustrated postcards and other wares, noting that "of all other lines the bicycle probably needs the least introduction," implying that many talking machine dealers had once specialized in bicycles. The bicycle "craze" had reached its peak between 1894 and 1896, but fizzled away to almost nothing in 1897 due to overproduction and the unreasoned speculation of investors, both in the context of a trade characterized by seasonal highs and lows.[45]

The association of phonographs and these other retail goods unavoidably suggests a context for recorded sound. The seasonal equilibrium between bikes and phonographs, in particular, offers another reminder that such goods circulated amid a cultural economy in a modest sense determined by conversations about "New Women" and middle-class domesticity. Ellen Gruber Garvey (1996, chapter 4) has demonstrated persuasively the ways in which bicycles became the subjects and instruments of gender definition, according to which advertisers as well as the fiction editors of the monthly magazines represented women's bodies and helped to construct their roles as consumers. As Garvey explains, "safety" bicycles became defined by and within contemporary debates surrounding women's clothing, mobility, and sexuality. That the spasm of controversy over bloomer-clad, bike-riding New Women did not last long only suggests that the fillips of cultural panic that phonograph records would later help to stimulate were in some sense familiar ones. As collaborators in the foxtrot "craze" and later as vehicles for jazz, home phonographs and phonograph records helped to broach related questions about heterosociability and social dancing as well as youth culture and interracial mingling as aspects of public life in the United States.

In this chapter, I have been suggesting that "inventing" or "producing" recorded sound cannot be narrowed to the activities of Edison and Berliner, or the efforts of corporate entities invested in the manufacture, advertisement, or sale of phonographs and records at the turn of the twentieth century. To my mind, the medium of recorded sound provides an exemplary instance of cultural production snatched from the hands of putative producing agents. Even though it would remain largely in the grip of corporate interests (eventually multimedia-multinationals like RCA/Victor), and even as it contributed to the redefinition of "merely" listening as a form of musical participation, this was a medium deeply defined by users and the changing conditions of use.[46] Understanding its social construction suggestively complicates received notions of new media as the purposeful outcomes of corporate strategizing at the same time that it provides an opportunity to think a little more broadly about the tangled intertexts of mass culture: the glossy magazines, retail outlets, and national brands that informed consumption at the end of the nineteenth century and that continue to do so today. In this light, casting the emergence of mass culture as the shift from a tactile, craft-oriented world to a visual, mass-product one seems simplistic at best. Cultural history must also include the squeaks and noises of change. Historians and critics must be prepared to explain the intensity of modern, mass-cultural experiences as well as their extensive range and appeal.

A bit like newspapers or photographs and other print media, or the broadcast media that soon followed, phonographs relied on a logic of transparency, of pure mediation, that was as chimerical as it was an accessory to the imagination of self and community, to a sense of location amid social spaces and forces. As much as some of their promoters seemed to invoke the possibility, records could never be transparent windows between musical experiences at the concert hall and in the home. There were differences in sound quality, of course: the lacking aura of performative origination; differing somatic, emotional, and commercial investments; and differences in arrangement, instrumentation, and so on. Along with a certain amount of knowing, commercial participation (that is, along with paying), playing phonograph records involved additional, tacit participation in all of the emergent conventions of sound recording as a medium—among them things like the sound quality, duration, materials, functions, and subjects of recording. Even many people who didn't like or play records participated in knowing the medium, as Sousa did, for example. Such tacit participation in its semiritualized character forms that part of media and mediation that cobbles users into publics and narrower counterpublics alike, invisibly uniting "us" "all" as potential consumers.

In the case of early recorded sound, mediation seems clearly to have involved assumptions regarding women and their roles in society. It is not just that women were represented and reproduced on records, not just that they helped sell phonographs or appeared in advertisements; rather, it is that modern forms of mediation are in part *defined* by normative constructions of difference, whether gender, racial, or other versions of difference. Women's voices early provided a standard for both the desire and accomplishment of recorded sound. Gender colored distinctions between work and play, recording and playback, business and amusement. Gender infused contemporary experiences of reality and imitation, performance and mimicry, attention and distraction. And gender flavored the pursuits of middle-class self-improvement, self-control, and self-indulgence. Phonographs only "worked" when they got women's voices right, just as home phonographs only "worked" according to the ways they interlocked with existing tensions surrounding music and home, ongoing constructions of shopping as something women do, and the ways in which users of all sorts desired, heard, and played recorded sounds.

I know that this is at once taking a close-in view (looking for a "deep" definition of new media) and an arm's length perspective (looking for what Clifford Geertz has called a "thick" description of context). I have ranged from the nitty-gritty hardness of phonograph records to the broadest social contexts in which music and mimesis were experienced within the emerging consumer culture. Both levels are apposite. One might say by comparison that Janet Abbate's (1999) users of the ARPANET and early Internet must be located in both the technical specifications and material conditions of the digital network, and the broadest social contexts within which those specifications and conditions emerged. There is nothing incidental or merely happenstance, then, in the fact that *users* in that case refers largely to—as Mark Poster (2001, 37) puts it—male "individuals who happened to be graduate students." These users helped to define the new digital networks precisely as part of the culturally and historically specific experience of *happening to be* male graduate students in the years of the Vietnam War, the sexual revolution, the Twenty-sixth Amendment, and post-Sputnik Big Science. Their experience of distributed networking, with its incipient cyberspatiality, must have been profoundly informed by the prevailing conditions of academe, for one, including the patterns of federal research funding, emerging extraterritoriality of the research "multiversity," and contested spaces of U.S. campuses in the era of sit-ins, teach-ins, and other protests.[47] So too must experiences of recorded sound at the turn of the twentieth century have been profoundly informed by preexisting gendered contexts for mimesis, commodified leisure, and music.

Admittedly, part of the appeal of the term *users* to address and critique the more familiar producer/consumer distinction in this chapter is based on recent and somewhat romantic constructions of computer users. Not only have individual computer users—like the author of Napster or the scattered authors of open-source Linux, for instance—gained acclaim and in some cases notoriety for their programming interventions but appeals to general standards of "user-friendliness" have brought users sidelong into the public eye. I have no wish to further romanticize, and will return to ARPANET and

Internet users in part II below. But the present currency of the term *users* does offer an opportunity to speculate at the broadest comparative level. Like the "convenience" of recorded sound, the term *user friendly* is a way to mystify as much as refer to the ongoing and medium-specific dynamic between users and publics. I have been suggesting more generally that media and their publics coevolve, and that one of the evolutionary forces at stake might best be described as a sociological tension between users and publics, where publics are comprised of users, but where users are not always constitutive members of the public sphere. So one might speculate that just as recorded sound and other contemporary media became defined amid and as part of the late nineteenth-century and early twentieth-century tensions surrounding the role of women and other "others" in U.S. society, so in the late 1960s and 1970s distributed digital networks and the computing practices associated with them were party to kindred, if also different, tensions surrounding the role of young people and other "others" in U.S. public life.

The next chapter shifts the grounds of my analysis from the social meanings of media at the turn of the twentieth century to the social meanings of media during the postwar and cold war period. Just as the introduction of recorded sound in 1878 devolved on experiences of and engagements with writing, print media, and public speech, so the development and introduction of distributed digital networks in the United States depended on existing, if dynamic, contexts that had little on their face to do with digital communications. As different as phonograph records and the Internet certainly are, these media both emerged from and as part of broadly shared questions about what can and should be inscribed as well as much more tacit questions of inscribability and inscriptiveness that shot—and keep shooting—to the heart of U.S. public life and public memory. Unlike the early history of recorded sound, that of digital networks directly involved the state. And unlike the once startling power to capture, to materialize and differently commodify sound, what often seems so startling about digitization and distributed networks is their supposed power to *de*materialize and differently commodify information. But like invisibility, dematerialization exists only in keeping with its opposite. Any putative dematerialization, that is, can only be experienced in relation to a preexisting sense of matter and materialization, which is why the next chapter begins with the familiar materiality of paper cards and state bureaucracy.

II *The Question of the Web*

3 New Media Bodies

Card Stock

In its 1968 decision *United States v. O'Brien,* the U.S. Supreme Court tangled with free speech as a question of bibliography. The case concerned a young man in Boston who had publicly burned his draft card, and the Court was asked to decide whether his act might not be constitutionally protected under the First Amendment. The Earl Warren Court ruled against David O'Brien seven to one, and in the course of his decision, the chief justice chided the man for having been "unrealistic" in his characterization of draft cards as "so many pieces of paper" intended to notify draftees of their registration and then "retained or tossed in the wastebasket according to the convenience or taste of the registrant." These were not ephemeral notifications, not mere messages; they were lasting and indelible certificates, made so by a legitimate act of Congress. The point of the Selective Service card, the Court reasoned, was not just what the card said but also, more particularly, *that* it said: the physical integrity of this small white card made a difference because only as a physical (that is, bibliographic) body could it reliably facilitate "the smooth functioning of the system" of which it was part. That system, the Selective Service System (SSS), matched bibliographic bodies one-to-one with eighteen-year-old male human bodies in a highly rationalized way.

The Court's reasoning was tortured in several respects. On First Amendment grounds, the line the Court drew between speech and nonspeech has proved impossible to maintain in subsequent rulings on related cases.[1] On bibliographic grounds, the Court pointed to a number of different supposed functions that "would be defeated" by the destruction or mutilation of the cards. If draft cards were destroyed, they could not serve "as proof," Warren maintained, in a way that was "easy and painless" for draftees as well as "just and efficient" from the perspective of the SSS. As such, the cards worked like "receipts" that

Figures 3.1a and 3.1b *United States v. O'Brien* exhibits 1a and 2a: draft-card burning and the "charred remains," photographed by the FBI, March 31, 1966. (*Source:* Library of Congress.)

could certify a draftee's status in a "rapid and uncomplicated" manner by keeping necessary information handy. Further, if the cards were destroyed, they could not offer the "continual reminders" that they did, alerting registrants that they should be in touch with their draft boards about changes in status or address. And finally, if the cards were destroyed, it would "obviously" be much more difficult to identify abuses like alteration or forgery, which were clearly against the law.

In calling the Court's interest "bibliographic," I mean to underscore the complexity of its investment in the draft cards as meaningful, paper-based, textual forms. As texts, draft cards seem to have inhabited a vast and murky middle ground between two poles: the idea

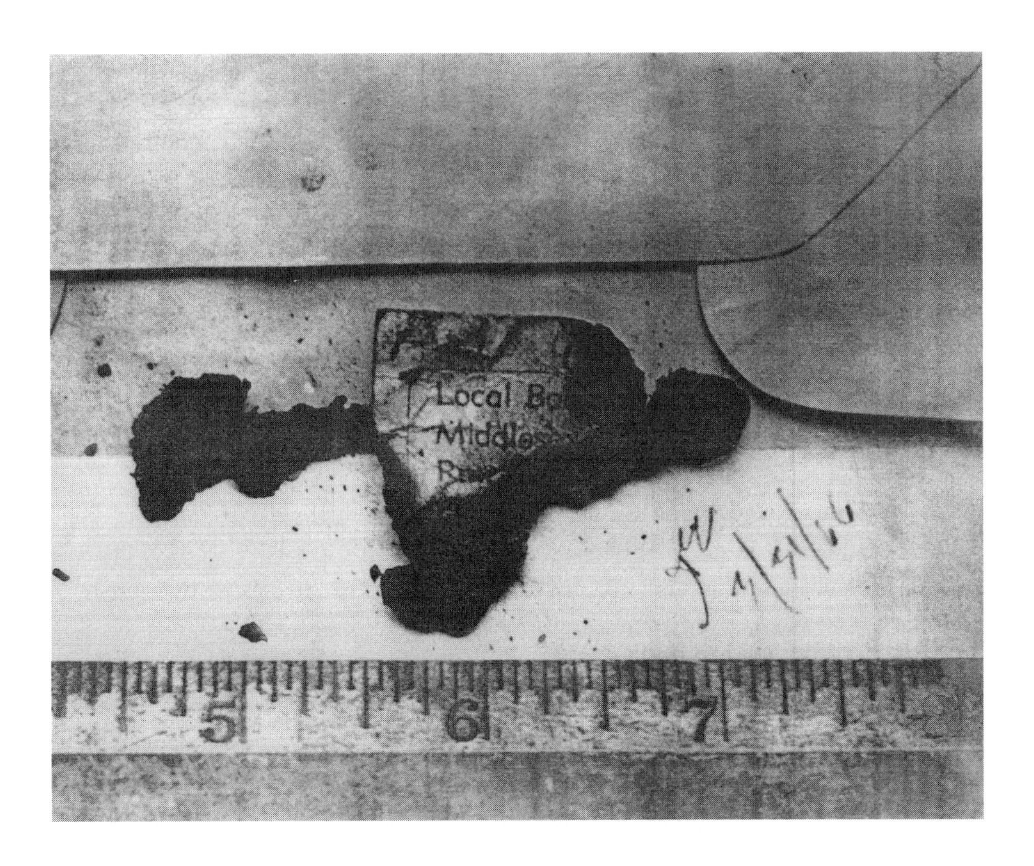

of a pure text, on the one hand, and some sort of nontext, on the other. O'Brien had opted for the pure-text extreme, according to Chief Justice Warren, when he argued that the cards were only meaningful for—only meaningful *as*—the information they contained. The opposite extreme would be an untext of some kind, like a bookmark, which contains no information itself, but functions *in its body* as a meaningful instrument.[2] At one pole there would be only meaning—unadulterated information, content, message, data—and at the other there would be only matter—an empty piece of paper, a blank, performing its function. Everything in between the two extremes is sign plus supplement, meaning and/as material. (That ugly "and/as" is necessary because meaning and materiality are mutual and not distinct.) Some of the murkiness of the middle ground between the two extremes is suggested in part by the Court's weird elaborations, such as the implication that destroying the cards might abet forgers of them. Its vastness is indicated by the presence of so many other denizens in the same middle ground: inscriptive media all inhabit that turf, including plenty of forms that are even less familiarly "media" than the quartos and folios that concern traditional bibliography. A ten dollar bill is neither pure text nor untext, for instance, since its meaning (its value) as ten dollars arises amid consensual circulations of related tender: a bill is not "only" its value but is also not "only" a piece of paper.[3]

While bankers and computer engineers pondered the meaning and/as material of checks, computer punch cards broadly raised many of the same questions that draft cards did in the mid- and late 1960s.[4] Both kinds of cards contained a variety of information and both were also instrumental to the systems of which they were a part. If punch cards seem more literally instrumental—more on the bookmark end of things—than draft cards do, ironically punch cards also seem more pure, or self-identical with the information they contain. The contradiction arises because the systems that used them were big pieces of machinery and so much of the information they contained could only be "read" by those machines. The cards might have columns, numbers, words, or icons printed on them for human readers, but their primary instrumentality had to do with their mysterious (that is, illegible) patterns of holes.[5] While it is unlikely that anyone asked at the time how punch cards worked like texts, using them raised this question by proxy.[6] Users had to know what code the cards required, and how to punch them or have them punched to contain programs and data. They had to know up from down, front from back, used from unused, correct from incorrect—all questions that concerned the cards as meaningful instruments and the patterns of holes punched in them as a form of—or a form resembling—semiotic presence.

Both kinds of cards also came with injunctions against destruction, made explicit by an act of Congress in the case of draft cards, and made manifest by printed warnings in the case of many punch cards. "Do not fold, spindle, or mutilate": These exact words may not always have appeared but they enjoyed currency as a catchphrase, part of the social meanings of computer technology and its applications.[7] Several months before O'Brien burned his draft card in Boston, the student newspaper in Berkeley joked that "lesson one" for every new student at the University of California was "not to fold, spindle, or mutilate his IBM [student registration] card." Another system meant to pair bibliographic bodies and eighteen-year-old human ones, but that semester the Free Speech Movement effectively turned the university's punch cards into tokens of authority, to be rebelled against, altered, and subverted: one card puncher was used to make holes that spelled out "FSM" and "STRIKE."[8] Like O'Brien burning his draft card, these radicals subverted bibliographic norms and expectations as an exercise in free speech. They took cards that were intended to work in one way and made them work in another.

Many people—on the Left, at least—simply stopped drawing distinctions between one card-enabled system and another. Whether the cards registered draftees or pupils, they helped "the system." Lewis Mumford (1970, 183) called it "the Pentagon of Power," and reflected that "today the increasing number of mass protests, sit-downs, and riots—physical acts rather than words—may be interpreted as an attempt to break through the automatic insulation of the megamachine, with its tendency to cover its own errors, to refuse unwelcome messages, or to block transmission of information damaging to the system itself." Mumford's metaphoric megasystem suggests the broad cultural currency that "cybernetic anxiety" had come to enjoy by the late 1960s.[9] Anxiety, like alienation, was one response to dehumanizing bureaucracies with dehumanizing machines. Digital systems in particular seemed poised not merely to displace humans—as industrial automation kept doing—but finally to erase the very distinction between human and machine. This was the slippery slope toward "the posthuman" condition, as it has been called by N. Katherine Hayles.

The point of these examples from the 1960s is to suggest that bibliographic questions persist in unlikely places and can have unusually broad implications. Arguments over both draft and computer cards revealed unsettled and yet widely held assumptions about the ways that meaning is authored as well as conveyed on paper. These assumptions served to underpin the normal (that is, the systemic) uses of the cards, and consequently to circumscribe their adaptive reuse or destruction as subversive. Yet because they remained unsettled, the same assumptions helped form ripples in the social order, bubbling to the surface in contests to define the contours of the U.S. public weal, in arguments over free speech and attempts to parse the difference between speech and nonspeech, between Mumford's "physical acts" and "words."

One of the core propositions of the preceding chapters has been that unsettled assumptions like these become more unsettled, or at least more *evidently* unsettled, by new media. Put another way, new media can be potent, embodied versions of unsettlement. To be sure, there was nothing new about draft or, strictly speaking, punch cards in the 1960s. The most celebrated new medium of the decade was color television, the commodity's commodity. Color broadcasting systems had been developed in the 1940s and 1950s, but RCA publicly launched its system with great fanfare at the 1964–1965 New York world's fair.

At first glance, television in black and white or color has nothing whatsoever to do with bibliography since television is noninscriptive: a television broadcast has no body—unless or until it is taped. But as Philip Rosen (1994, 225–234) has shown in his analysis of NBC's coverage of John F. Kennedy's assassination, part of the defining portentousness of television arises from its use of inscriptions. Part of the anchorperson's presentation of that historic event as both self-evidently historic and uniquely available to viewers was a tension between the anchors' live narration and the still photographs they held up to the camera, the wire-service reports they read, the audiotapes they played, and particularly—more than two hours after the event—the first in-the-can footage from Dallas, taken earlier that November afternoon in 1963. The live broadcast was chock-full of inscriptions. Rosen calls this the "subtextual drama of the medium's struggle to depict itself," which was played out in the newscasters' and the network's evident discomfort at "the structuring absence of images from the key scenes of the motivating action" (228, 229). Even this preeminent noninscribed media event, the emergency broadcast, must be involved with and defined by its manipulation of as well as its partial yet repeated distinction from inscriptive forms.

I want to be clear that the media of the 1960s, new or old, in this respect have much in common with the new medium of recorded sound that I have described at length. I argued most explicitly in chapter 1 that when recorded sound was new, it was in some ways experienced as party to the existing, dynamic logics of writing, print media, and public speech, the nexus of so many open questions I have here called bibliographic ones, because I started with meanings authored and conveyed on paper. The new medium came to make sense only when its demonstration to and subsequent use by early audiences helped to construct a coincident yet partly contravening logic for recording—a logic that soon became self-evident, and thus came to seem intrinsic to phonographs and phonograph records. Furthermore, I suggested that the implications of the emergent logic for recorded sound extended far beyond the eventual formal conventions of the medium. Like the bibliographic questions that got so tangled up with the First Amendment in the 1960s, questions concerning the new and curious inscriptive qualities of sound recording were similarly entangled. Instead of free speech, recorded sound helped to broach questions about speech itself, the means and meaning of its selective preservation, which in turn helped to broach questions about the scope and character of U.S. public life and public memory.

This chapter concerns media that are more familiarly new than the new media of 1878–1910 or color television. I want to address digital networks as new media, both because they continue to dominate present thinking about media in general and newness in particular, and because they proffer an important instance of—very broadly—bibliographic questions and entanglements. Digital networks are many other things too, to be sure, and this chapter cannot presume to offer anything like a full accounting or complete history. My purpose instead is to strike a few notes of comparison between apples and oranges—between acoustic recordings and texts on the Internet—because I think those notes can lead, at least speculatively, to a better understanding of the ways that media accumulate the power that they do.

The differences between early phonograph records and the Internet are as obvious as they are undeniable: analog and digital, mechanical and electronic, consumer good and communications medium. Nevertheless, as new forms of inscription these two new media shared similarities. If sound recording helped to call the mutual meanings of print and public speech into question, electronic documents have raised questions of a related sort. Distributed digital networks and the texts they make possible have emerged amid an existing textual economy, a world and workplace powerfully self-constituted according to the logic of contemporary media: print publication, broadcasting, Hollywood, and the record labels, but also punch cards, printouts, and paperwork. Experiences with digital networks have helped to construct a coincident yet contravening logic for digital texts, partly in response to material features of the new medium, and partly in response to the hugely varied contexts of their ongoing reception and development. As digital media continue to change, the logic of electronic texts of course remains inchoate, although certain elements of that logic have already become conventional, and thus self-evident or transparent and difficult to see.

Like television, digital media at first blush have little to do with inscription. Inscriptions are both material and semiotic, and yet digital texts can seem strangely immaterial or disembodied. Like so much online, they are often thought of as "virtual" because they are so elusive as physical objects. No Web page would exist without a vast clutter of tangible stuff—the monitor on which it appears, but also the server computer, the client computer, the Internet "backbone," cables, routers, and switch hotels—but it is nonetheless strikingly intangible. What is it? Where is it? In this way, electronic texts seem to gesture toward the pure-text extreme dismissed as so unrealistic by Chief Justice Warren in relation to O'Brien's draft card: a digital text, according to this line of thinking, is only meaningful *as* the information it contains, so it is no problem—and no wonder—if the digital text seems to have no body.

Many scholars have recently pointed out the pitfalls of such reasoning, arguing for and exploring the varied materiality of digital texts. The best among them call for a "bibliographic/textual approach" to consider elements like "platform, interface, data standards, file formats, operating systems, versions and distributions of code, patches, ports, and so forth," because *"that's the stuff electronic texts are made of."*[10] But what sort of stuff is that? Even the most astute and exacting critics of cyberculture tend to signal a certain ambivalence about the bodies that electronic texts have, judging at least from the frequency with which the word *material* appears between scare quotes. Lev Manovich (2001, 45, 48) writes that the "basic, 'material' principles of new media [are] numeric coding and modular organization," and that hardware and software have "'material'" as well as "logical principles." Similarly, Mark Poster (2001, 77) chides that "the impact of technologies is never a linear result . . . of their internal, 'material,' capabilities."[11] The quotation marks around *material* serve obliquely to interrogate the claims being made. Both critics imply that there is no putting a finger exactly on the matter at hand, if logic is logic, but material is "material."

This is an ontological problem as much as a semantic one, a quandary over what a digital text fundamentally *is,* which has led, it turns out, to some productive wrangling over what a nondigital text fundamentally is by comparison. "What Is Text, Really?" Steven J. DeRose, David G. Durand, Elli Mylonas, and Allen H. Renear all wanted to know in the *Journal of Computing in Higher Education* (1990). "What is a text?" asked Espen J. Aarseth in *Cybertext: Perspectives on Ergodic Literature* (1997, 15) and Johanna Drucker in a review titled "Theory as Praxis: The Poetics of Electronic Textuality" (2002b, 685). "What is *text?*" Matthew Kirschenbaum insisted in *Unspun: Key Concepts for Understanding the World Wide Web* (2000, 127).[12] Critics always ask these questions rhetorically, in order to answer them, although their recurrence and persistence as questions in these contexts demonstrate an ongoing negotiation over the very subject of humanities computing. And in the much broader world of computing generally, similar negotiations abound. Any new editing, publishing, or scanning software, every new markup grammar, plug-in, or Web applet proposes a new idea of what electronic text actually is. The question is not asked as explicitly, but the answers are all there, and many of them are for sale. Any electronic text "embodies specific ideas of what is important in that text," and thus embodies specific ideas of what electronic text can be (Sperberg-McQueen 1991, 34). The high-stakes "drama of the medium's struggle to depict itself," as Rosen puts it, gets worked out both offscreen and on-screen, amid and as part of the political economies of knowledge work.[13]

Taking a page from DeRose et al. (1990), this chapter and the next explore digital text partly in relation to nondigital text. Just as the tinfoil records of 1878 and the wax cylinder records of 1889–1893 are knowable largely through the newsprint records that also helped to define them by contrast, electronic texts are knowable partly through and by contrast to "the social life of paper" and partly in relation to filmic, magnetic, and other inscriptions that are not paper based.[14] Just as chapter 1 looked at phonographs before they were even practical, everyday devices, this chapter considers digital text from the moment before digital networks were practical and familiar, as it appeared on the ARPANET, the precursor to the Internet (put together by the Advanced Research Projects Agency or ARPA) in 1968–1972. Structuring my inquiry this way allows me to sidestep the ontological problem of what electronic text *is* in order to try to understand how the *question* of what electronic text is ever occurred in the context of digital networks in the first place. What kinds of experiences might early users of digital networks have had that implicated those networks as relevant to issues of bibliography at all? What were the contexts of use and usability on the emerging network? Only by tackling these issues first will it be possible to tease out some of the sprawling implications that the relevant bibliographic questions may have. The following chapter takes a stab in the latter direction by focusing on the World Wide Web today.

Sources for my immediate project are admittedly scant. Electronic records from the ARPANET era have rarely survived or have been "ported forward" to more recent technology in such a way that their ARPA-ness, their telling archival character or provenance, whatever it might have been, has become obscure.[15] Published sources from the ARPANET era are also scarce, particularly when compared to the bubble of attention that Edison's first phonographs received. In the popular press, for example, the *New York Times* only mentioned the Internet once before 1988; then too, there was little relevant trade literature to speak of, and scholarly publication in the nascent field of computer science was only beginning to gain momentum.[16] As an agency within the U.S. Department of Defense, ARPA and its Information Processing Techniques Office (IPTO) didn't publicize its activities. Even the IPTO's contract bidding process was—an oxymoron—selectively public.[17] The dearth of electronic sources in particular presents a problem, although it also addresses one of the central points of my argument about new media and media history. Much of the potential standing that electronic texts have as evidence is an outcome, not a precondition, of the logic that digital media are coming to possess specifically in relation to writing, print, and other media.

Keyword: Document

When J. C. R. Licklider went to the Pentagon in fall 1962 to head APRA's new IPTO, he took a leave of absence from Bolt, Beranek, and Newman (BBN), a Cambridge, Massachusetts, firm of engineering consultants who specialized in acoustics. Over the next few years, Licklider would help to envision and initiate the ARPANET project, and then BBN would be hired under his successors at the Pentagon to build the network's core Interface Message Processors (IMP). In the meantime, one of the things Licklider left hanging at BBN was a consulting gig for the Council on Library Resources, Inc., funded by the Ford Foundation. Though removed from the day-to-day research, now directed by a colleague at BBN, Licklider kept track of the project and eventually authored its report. As the Council on Library Resources noted, "The 'research on concepts and problems of libraries of the future' continued under [Licklider's] general direction in his absence," while he was "on a special assignment for the Department of Defense." The council got its final report in January 1964, and The MIT Press published *Libraries of the Future* in 1965.[18]

The impetus for the libraries project arose from what Carolyn Marvin (1987, 59) has called a "digital" view of information, although the term *digital* was not used as freely at the time. The digital view assumes that "information is a definite quantity increasing daily," as opposed to what Marvin identifies as an "analog" view, characterized by the assumption that information is "a constantly shifting and repatterned feature of every environment, past or present, a transaction between knowers and what is known."[19] The digital view of information has Enlightenment roots, but it emerged with force in the post–World War II, post–Manhattan Project era, as part of a widely shared anxiety about the continued efficacy of science in U.S. life—an anxiety that only intensified with the Soviet launch of Sputnik in 1957. In particular, pundits worried, science faced an impending bibliometric crisis: too many publications. Increasing specialization and ever more research meant that scientists—and all of civilization by extension—were being buried alive by the accumulated mass of paper. If left unchecked, Mumford (1970, 182) summarized, the sheer "overproduction of books will bring about a state of intellectual enervation and depletion hardly to be distinguished from massive ignorance."[20] University and other research libraries were already "becoming choked from the proliferation," according to the Council on Library Resources (Licklider, v). So the council hired Licklider and BBN to work on the problem.

The most famous statement of the impending bibliometric crisis is Vannevar Bush's "As We May Think," which had appeared in the *Atlantic Monthly* in 1945.[21] The article offers a meditation on the future of compiling and consulting the huge mass of accumulated human knowledge, which Bush variously calls "the great record," "the total record," and "the common record" of humankind. Bush's formulation is famous because he describes a futuristic solution to the problem: a method of storing and sorting information that is modeled on human consciousness rather than bureaucratic filing systems. It is an imagined hypertext, before the term *hypertext* was coined and any of the relevant digital technology existed. He suggests that documents might best be organized not by "artificial" indexing systems with their rigid "paths" and cumbersome rules but by a more natural form of "associative indexing," working in the manner of the "intricate web of trails" that connects related thoughts in the brain. To this end, Bush posits a device, "a sort of mechanized private file and library" to serve as an "intimate supplement" to its user's memory. Dubbing it the *memex,* Bush envisions "a desk [that can be] operated at a distance," with screens on top and microfilm contents inside to be somehow selected, consulted, annotated, and then joined or tied at will into multiple associative "trails" for future reference.

Bush's presumptive "we"—"As *We* May Think"—like his preoccupation with "the common record," signals a media public of the sort I explored in chapter 1. Like the exhibitors of tinfoil phonographs, that is, Bush's memex posits the shared ownership of a public record, presumptively "owned" by "us" in the commonsensicality of its own production, preservation, and potential retrieval for a shared public good. Licklider tackled the libraries of the future project with all of these same assumptions in place. What he added was his own futurism, an exacting interest in "Man-Computer Symbiosis," and an extraordinarily astute analysis of libraries and their role in the production of knowledge.[22] Though he claimed not to have read Bush's "pioneer article" until completing his own work, Licklider acknowledged its indirect influence, "through the community," and dedicated *Libraries of the Future* to Bush (xiii). Licklider's volume is much less well-known than Bush's article, but it offers an opportunity to gauge the ways that computers and texts could be thought together in the moment just before digital networks became an accomplished fact as well as a new medium.

The future Licklider takes as his point of orientation is the year 2000, and the libraries he proposes are what he calls "procognitive systems." In order to envision these future procognitive systems, Licklider must reject many of the "schemata" by which traditional libraries are understood as well as many of the schemata by which computers were understood in the mid-1960s. He is in favor of the page, for instance, as a schema for the display of information ("As a medium . . . the printed page is superb" [4]), but dismayed by its "passiveness" as a means of long-term information storage. Books have even less to recommend them, and library stacks of books still less. Likewise, Licklider is in favor of some of the features of computing, like random-access memory (RAM), but not of the existing technological limitations or the mind-set that associates computing with writing a program and delivering a stack of punch "cards to a computer center in the morning, and picking up a pile of 'printouts' in the afternoon" (8). By carefully rethinking libraries and computers in such schematic terms, Licklider arrives at a wishful future in which researchers sit at consoles or terminals, typing on keyboards and looking at screens, connecting to and interacting with digital systems to query, search, and retrieve information.

Licklider's procognitive systems are amazingly prescient. Computing in "real time," when a user and computer interact through keyboards and screens, was all but impracticable in 1965, known primarily to those few who worked on advanced systems for the military. Drawing on his experiences with computing at MIT, where real-time and time-sharing (multiterminal) computing were both under continuous development, Licklider had settled on the same man-at-his-desk schema that had shaped Bush's memex and that would so influence the graphic user interface—with its metaphoric desktop—of personal computing more than a decade later.[23] Licklider envisions a year 2000 in which "the average man" owns or rents his own terminal and uses it to connect to a network. "In business, government, and education," he speculates, "the concept of 'desk' may have changed from passive to active: a desk may be primarily a display-and-control station in a telecommunication-telecomputation system—and its most vital part may be the cable ('umbilical cord') that connects it, via a wall socket, into the procognitive utility net" (33).

The image of man and his umbilical cord is a striking one, but what kind of actual or bibliographic body will "the body of recorded knowledge" (6) have when Licklider's procognitive systems become the libraries of the future? Licklider and his research team recognize some limits: "We delimited the scope of the study, almost at the outset, to functions, classes of information, and domains of knowledge in which the items of basic interest are not the print or paper, and not the words and sentences themselves—but the facts, concepts, principles, and ideas that lie behind the visible and tangible aspects of documents." By focusing on information that exists "behind" the visible and the tangible, Licklider and his team agree to set aside all works of art, graphic and literary. Instead, they focus on cases where they suppose that information can be separated from its natal body "without significant loss": both art and literary criticism, perhaps, as well as "most of history, medicine, and law, and almost all of science, technology, and the records of business and government" (2). The information "corpus" so disassociated from pages, books, and shelves would be introduced to a new body, an advanced, "processible" memory system or binary storage medium, as yet undeveloped. That memory, Licklider estimates, would have to have at least the capacity for 5×10^{15} bits, where each alphanumeric character in the corpus is represented (conservatively) by 5 bits, taking 5 cells of binary storage space (15–20, 63–64).

Without a specified core memory, Licklider's procognitive systems lean toward the pure text ideal hinted at in the O'Brien decision, with one notable addition. Licklider compares his procognitive systems to a world in which a "document [can] read its own print" (5–6). The systems would be able to answer questions as well as retrieve documents, because they would "be able to 'read' and 'comprehend' the documents themselves" as well as the tags or catalog information associated with them (153). So Licklider's electronic documents are self-identical and self-reading. They *are* the information they contain; yet they also consist recursively of representations, "smart" versions of themselves and the paper documents they reduce to facts, concepts, principles, and ideas.

By *document* Licklider explains that he means a document "type," not what he distinguishes as individual document "tokens." A document token is a unique and physical copy of a document, whereas a document type is a whole edition of like tokens from which any representative may be selected to have its alphanumeric characters spelled out into the system (13). Any variation among document tokens is, like typography, assumed to be trivial, dispensable, or what bibliographers call "accidental" or nonsubstantive. But if textual variants among tokens are unimportant, Licklider still wants his procognitive systems to retain a number of concepts or schemata relating to variations among document types, like genre: monographs are different from articles are different from reviews. And he wants to retain schemata relating to the hierarchical parts that document types have, such as sentence, paragraph, chapter, and volume as well as author, title, abstract, body, and footnote (7).

Libraries of the Future contains a brief survey of the relevant research on information storage, organization, and retrieval along with a section on the linguistic analysis that might be required in the "eventual machine 'understanding,' of natural-language text" (131). Licklider himself envisions a sequence in which the "natural," technical language of science would undergo "a machine-aided editorial translation" into an unambiguous or "ruly" English, which would in turn undergo "a purely machine transformation" (that is, scanning) into the "language(s) of the computer or of the data base itself" (89). For the present, that means that the proposed transmigration of information from natal to self-reading body hinges on the conversion of documents into so many alphanumeric character strings (as well as some sort of unspecified "adjustment" for pictures and other nonalphanumeric contents) that are retrievable and readable by as well as via the procognitive systems. Alphanumeric characters are essential, but actual letterforms don't matter, even though ironically the document at hand, *Libraries of the Future,* depends on font changes to signal shifts between Licklider's analysis and the examples he gives of hypothetical interactions with both a local computing system and the procognitive one. Each system is represented by a different font, and both are different from the prevailing typeface of the book. "Are you J. C. R. Licklider?" the local computing utility asks Licklider in a typewriter font. "I type 'y' for yes," Licklider narrates entirely in the dominant font of the book. When he types "Procog" to log on, the system responds, "You are now in the Procognitive System" in a small san serif typeface (47–48).

Licklider gives an illustration of a procognitive system in use that is strangely ambivalent in its futurism. He uses a procognitive system from the future, but to pursue a question he has in the present, explicitly as if he had "available in 1964 the procognitive system of 1994." He asks the system about "the prospect that digital computers can be programmed in such a way as to 'understand' passages of natural language" (46). The procognitive system searches ten thousand documents for relevant material, then presents Licklider with citations and abstracts, and has full-text versions "available in secondary memory," if he wants them. His online session is dated 14:23 November 13, 1964, and the bibliographic citations he retrieves refer to articles from 1961 and 1963 that appear in his own bibliography. In this respect, even though the paper-based *Libraries of the Future* is not a digital or self-reading document, it is represented by Licklider's example as a self-writing one. The conceit of self-writing whereby *Libraries of the Future* imagines the future research that leads to *Libraries of the Future* offers an echo of the recursive logic that any system of self-reading documents must possess as well as the reflexivity or self-study that would be offered two years later as one of numerous reasons for building the ARPANET.[24]

Many aspects of Licklider's future do resonate with actual developments in computing over the decades that ensued. If the desk schema, query, search, and network interactions all make sense to today's computer users, many of Licklider's more specific ideas about digital texts also make sense in hindsight according to later developments. Licklider's wishful 5-bit alphanumeric code can be recognized in the 8-bit ASCII standard adopted by the ARPANET in 1969. His wishful commitment to genre can be recognized in today's document-type definitions or markup grammars, which text encoders construct for different classes of documents. And his wishful commitment to the hierarchical parts that documents have can be recognized in the principle of defining texts as ordered hierarchies of content objects on which markup strategies are premised.[25] Taking *Libraries of the Future* on its own circa 1965 terms, however, helps to index more about digital inscriptions than the forms they have recently taken or the theorizations they may lately have provoked.

In particular, Licklider's volume demonstrates both the pertinence and complexity of the term *document*. Like the related term *record* in 1878, the term *document* in the 1960s was nothing new. As a noun, it hails in its present form from the eighteenth century, when, according to the Oxford English Dictionary, it referred to "something written, inscribed, etc., which furnishes evidence or information upon any subject, as a manuscript, title-deed, tomb-stone, coin, picture, etc."[26] As Licklider uses the term, documents have something of a synecdochic function: documents are not unique artifacts (his "tokens"); they are representative ones. In addition to furnishing evidence or information on their subjects or contents, they furnish as well as delimit evidence about the whole edition they represent. (They importantly delimit evidence because they are based on assumptions about which details matter and which do not.) Moreover, since they are called "documents" when they are shelved on a bookcase and also once they are entered into the procognitive system, their status is doubly synecdochic online: electronic documents are representative-substantive copies, simulacra, of already representative-substantive forms. They furnish as well as delimit evidence about types that furnish as well as delimit evidence about tokens.

Documents in *Libraries of the Future* may be digitized, but they cannot be de novo digital creations. Licklider proceeds according to an unstated distinction between documents and what Matthew Kirschenbaum (2002) has called "first-generation electronic objects." Licklider's electronic documents are third-generation, edited from natural language into "ruly" language and then processed into machine code. In Licklider's volume, first-generation electronic objects appear only as the prompts, headers, queries, results, and other online interactions that are represented on the procognitive screen or—one of the hardiest schemata of computing in 1965—as printouts on sheaves of paper. Put another way, there is always an inside and an outside to documents within the Licklider system. Documents are distinguished from programs as well as "raw data, digested data, [and] data about the location of data" (Licklider 1990, 29).

Today, the meaning of *document* is much broader than this and at least as complex, to judge merely from the "My Documents" folder on any personal computer running Windows, the ".doc" files created and accumulated by Microsoft Word, or the sophisticated software packages, hardware alternatives, and consulting services available for document management.[27] Documents today may be created digitally, not just digitized. Licklider's

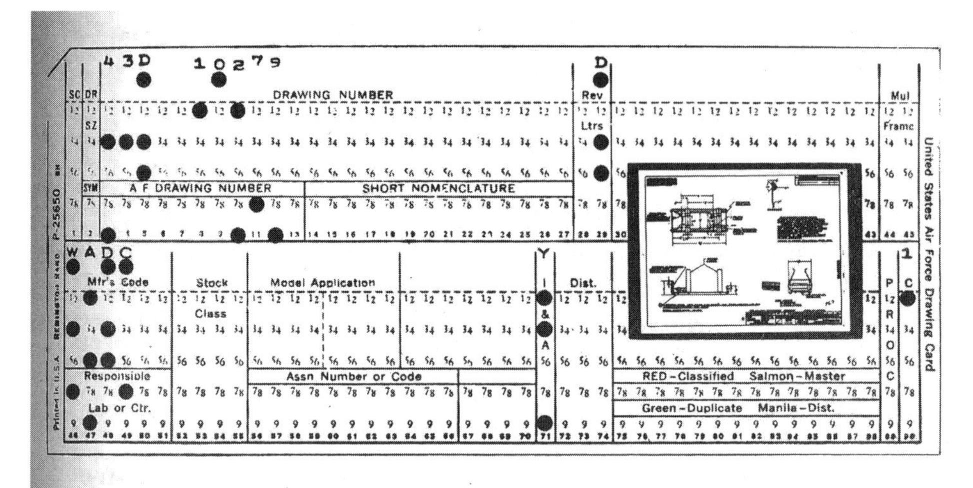

Figure 1. Remington Rand Format

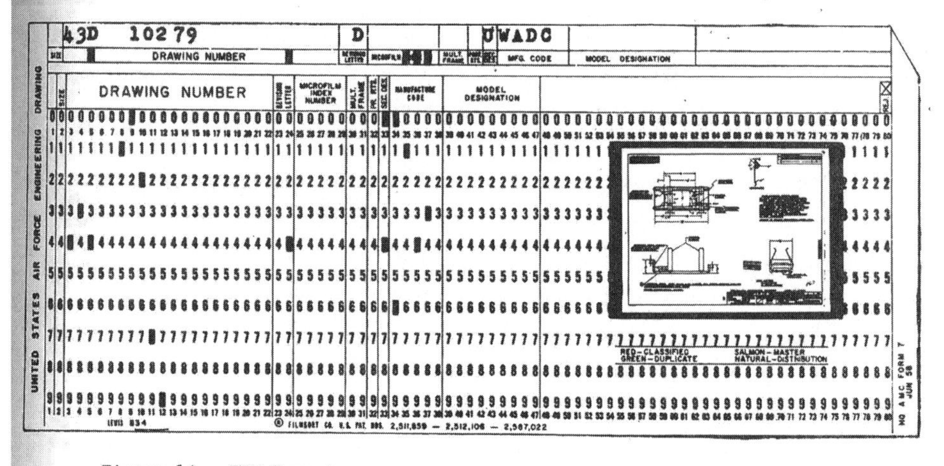

Figure 1A. IBM Format

Figures 3.2a and 3.2b Microfilm aperture cards and reader, 1960. (*Source:* Charles Babbage Institute, University of Minnesota.)

usage still lingers, however. It helps to explain why the Xerox Corporation trademarked the slogans "We Document the World" (1988) and "The Document Company" (1991), with the latter mark still in use. Indeed, Licklider's electronic documents in some respects resemble "smart" or self-reading photocopies. Like a photocopy, his digitized document furnishes as well as delimits evidence of the preexisting paper document it represents. Likewise, just as there is no de novo digitized document in Licklider's scheme, there can be no de novo Xerox, a photocopy made wholly without reference to a preexisting document. The Xerox Corporation's slogans, of course, have much to do with its long-lived ambitions to move beyond photocopiers and toward the synergies of the modern office environment. As early as 1966, the company needed engineers and other specialists to help develop "document management and computer-related systems"; help-wanted advertisements started to include "document management" as a discrete object for R & D around that time.[28] Among the most inventive systems for document management (not by Xerox) were the microfilm aperture card systems, a primitive database of images controlled on computer punch cards and read on a memex-looking desk.[29]

The present meanings of *document,* to stay briefly at this suggestive, semantic level, are no more singularly the result of digital media than the enlarged meaning of the word *record* in 1878 was the singular result of Edison's invention. Phonograph records added one specific meaning to a word already in the process of redefinition; they emerged from and engaged a conceptual field already subject to stress. Newly applied to tinfoil sheets as well as differently and newly applied to people (the politician's record, the ballplayer's record, and so on) the word *record* as it was used around 1878 hinted at what seemed important to record and important about recording, while both forms of importance appeared keenly in question amid the varied cultural politics of the post-Reconstruction era. The term carried with it and was reflexively defined by that portentous sense of abstract public or "common record" to which Bush later appealed in his imagination of the memex, although the public of 1878 was hardly more certain in its extent than that of 1945 or of today. Likewise, digital media have added specific—though notably loose—meanings to the term *document* as part of an already ongoing redefinition that remains complicated by and within a sense of abstract public and common record, of what seems important to document and important about documents.

"Importance" in this context alludes to that nexus of cultural values attending and reflexively defined by overlapping and dynamic media publics. Those values may be glimpsed in other uses of the term *document,* as in such varied productions as "documentary" editions and films. Documentary editions are the work of historians and textual critics who locate, authenticate, select, annotate, transcribe, and publish documents for use by students and scholars in the future; for example, *The Papers of Thomas A. Edison, The Margaret Sanger Papers,* or *The Mark Twain Papers.* They are documentary in the sense that they assume the merits and efficacy of historical and textual editing as a professional practice, of the work that humanists do by reading, assessing, and citing primary source material to support narratives about the past. Documentary editions supplement the "database logic" of archives, Lev Manovich (2001, 218–232) might say, promoting documents for their explanatory power (Manovich's "narrative logic") according to the norms of the historical and literary-critical professions in addition to the broader cultural economy that supports those norms in their varied attentions to Edison, Sanger, Twain, and the like. Documentary films by contrast are documentary in the sense that they assume the merits and efficacy of the filmic medium to represent reality or present nonfiction. Both uses of *documentary,* like so many uses of *document* as a noun and verb, differently denote a connection between texts and facts, where the facticity of texts inheres in their varied self-evidence as meaningful bodies that are also potentially authentic, original, unique, complete, uncorrupted, lasting, immutable, citable, or otherwise "true."

The related genealogies of *record* and *document* suggested here together point toward a history of the facticity of the modern text. Where has the present, if varied, self-evidence of texts come from? How have the meanings of textual bodies evolved as such, and according to which culturally and media-specific conditions? What would the history of bibliography look like in the broadest possible terms? Such a project would be in keeping with the "history of the modern fact" that Mary Poovey (1998) has discerned first in the construction of numerical certitude by Renaissance bookkeepers, as well as with "the invention of communication" that Armand Mattelart (1996) has discerned first in the work of seventeenth-century topographers.[30] It would have to include—even more self-reflexively—the genealogy of *text* in the postmodern or "antidisciplinary" sense that John Mowitt (1994) has dated to the poststructuralist movements of the same late 1960s that have concerned me here. And it have to would include an ample share of what is now called "material culture studies," a genealogy of cultural things and thingness.

One might begin this history of the modern text with the conceptual emergence of editions, and thence of modern bibliography, whether from the *pecia* system among Renaissance scribes (a division of labor among copyists) or the print shops of Gutenberg and his contemporaries. Or one might begin this history with the conceptual emergence of print as a fixed, authorizing, and authored medium, as with the British Stationers' Company and the Royal Society.[31] There are many other alternatives too. Wherever and whenever one begins, the printed book, like that later construction, "literature," is only one of its themes. Such a history must include the history of the footnote and photography, the imagination of the memex, the burning of draft cards, and the birth of digitization.[32] The varied means and subjects of inscription all offer themes, as do the varied political and cultural economies of making inscriptions public and consuming them. This whole history of the modern text cannot be my purpose here, yet it does offer a reminder that the histories of sound recording and electronic documents go together as one, not only as two.

1968–1972: Networking

The early architects of digital networks had to reckon with bibliographic questions on a number of different levels. The first, most obvious level was an immediate condition of bureaucracy. The ARPANET began with four nodes located at four universities that were already research contractors for ARPA. Representatives from each of the first four sites met during 1968 to begin planning the network, and the minutes of these early meetings were copied and circulated to ARPA. Primary investigators brought and later sent their graduate students to these Network Working Group (NWG) meetings, and representatives from BBN, the contractor selected in January 1969 to provide the interface computers at each node, joined them in February. It wasn't until March 1969, apparently, that it occurred to members of the resulting and somewhat ad hoc group that they themselves should really begin to write things down. This is not to say that paperwork had not already been accumulating at precipitous rates, particularly in Washington, DC, just that the group itself, this nascent online community comprised largely of graduate students, had taken few pains nor realized the need to generate a collective memory of its own over which its members—or the Department of Defense, or anyone—might have bibliographic control.

At the same time that it picked BBN to start construction of the IMP, ARPA had also awarded the Stanford Research Institute (SRI) node a contract to become the "Network Information Center" (NIC), a documentation site for the projected network. The resulting and initial efforts to document the in-progress ARPANET, by both the NWG members and the nascent NIC, together demonstrate an emergent interest in the genres and textual or bibliographic bodies appropriate to the new medium, an interest that would form the basis for an emerging "recursive public," in Christopher Kelty's terms (2005, 186), "a particular form of social imaginary through which [a] group imagines in common the means of [its] own association."

In April 1969, the NWG announced a new genre for the new medium: the request for comment (RFC). The name was an appropriation of the Pentagon's habitual requests for proposals, or RFPs, which solicited proposals as part of its contract bidding process. Put together, RFCs were "a set of notes" in which the NWG discussed the requirements for the network and established its initial protocols. There had to be a protocol for establishing protocols, however, so graduate student Steve Crocker tried to define the RFC genre in RFC 3 (April 9, 1969): "The basic ground rules were that anyone could say anything and that nothing was official." In that memo, he specifically encourages "philosophical positions without examples or other specifics, specific suggestions or implementation techniques without introductory background explication, and explicit questions." As Crocker explains, "These standards (or lack of them) are stated explicitly for two reasons. First, there is a tendency to view a written statement as ipso facto authoritative, and we hope to promote the exchange and discussion of considerably less than authoritative ideas. Second, there is a natural hesitancy to publish something unpolished, and we hope to ease this inhibition." RFCs are writing that seeks to be less written, publications that seek to be less published, and records that might be less than permanent. Crocker apparently believed that RFCs were "temporary and the entire series would die off in a year or so once the network was running." He thought that the online community (his "we") would eventually have the networking protocols established by the RFCs, but it would have no collective memory of its own apart from their implementation.[33] For the sake of clarity, Crocker proposed, "Every NWG note should bear the following information" in its heading:

"Network Working Group"
"Request for Comments:" x,
where x is a serial number
Serial numbers are assigned by Bill Duvall at SRI
Author and affiliation
Date
Title. The title need not be unique.

If Crocker was imagining a new form of document, he was still constrained by the existing technology. Originally RFCs were memos distributed on paper. At the outset, six copies went by mail, one each to Washington, the four nodes, and BBN, where further copies could go by hand. Within months, RFC 10 (July 29, 1969) updated the distribution from six to nine paper copies and gave mailing addresses. Subsequent RFCs enlarged the mailing list still further, and refined the processes for numbering and issuing them. RFCs circulated the parameters as well as the contested protocols of circulability, their own circulability as texts as well as the future circulability of packets, the unit of data transmission across so-called packet-switched networks like the ARPANET and its successors. As Kelty puts it, standards and protocols were "bootstrapped, first through a process of writing RFCs, followed by a process of creating implementations that adhere loosely to the rules in the RFC, then observing the progress of implementations (where they differ from the RFC, how many people are using this implementation, and other less obvious criteria), and then rewriting the RFCs so that the process begins all over again" (2005, 198). When early RFCs were composed and stored online, like RFC 2, they still had to be printed and distributed on paper, because the digital network as such did not yet exist.[34] Even after the first nodes were connected, when the NIC made its first important plug for online, networked documentation, it did so in terms that stressed the importance of paper: "Particularly useful would be on-line documentation. It is suggested that each site have an identical hard copy device (e.g., a line printer) suitable for reproducing documents."[35] Online documentation was only going to work if every site could download to paper. Like draft cards, documentation was for keeps, and in 1970 keeping still suggested "hard" paper copies—the NIC's printed "reproductions"—stored in local files.

The RFCs work as a form of documentation for the network and thus as a performative self-iteration of themselves as networked information. (The series continues today.) Some RFCs stand as protocols, rules for building and using the network—they articulate Telnet, FTP, and so on; others simply document the refinement, existence, and accumulation of standards and protocols. In February 1971, RFC 100 was issued as "a guide" to the series as it existed then. Its author, Peggy Karp, categorized the RFCs by subject, indicating "whether the notes are current, obsolete, or superseded," and "for historical reasons," gives "a brief summary" of each RFC "relevant to a particular category." Further efforts to control the accumulating information followed.[36] Eventually, RFCs would be published and stored as public access files in ASCII, and an electronic message was then sent around letting subscribers on the network know that they were ready for download. After more than a decade, RFC 825 (November 1982) would explicitly formalize the "rules" for RFCs that had emerged as a result of "the constraints of a wide variety of printing and display equipment" across the network. If RFCs were going to be accessible to all, they had to be in ASCII; they had to consist of fifty-eight-line pages made of seventy-two-character lines, and "No overstriking (or underlining) is allowed."

Despite formal properties at least as rigid as those that define a sonnet or a villanelle, the RFC has been celebrated as an inherently democratic form. Like e-mail in 1972 and Usenet groups in 1979, that is, the RFCs are part of an already often-told history of the Internet that emphasizes the network's grass roots, its community-based self-definition of the communication protocols that include multiple layers of technical specifications, but that extend as well to the conventions and popular accessibility of online discourse.[37] While in one sense there is clearly something tendentious about describing any element of an insular, "closed world," Pentagon-sponsored, R & D project as "democratic," in another sense there is also something a little banal about it in this instance, since all genres, as genres, are socially derived.[38] What remains critical to note is the way that ARPANET standards and protocols emerged in interlocking layers. The layer at which the RFC became defined as an ASCII file of a certain kind was in part a function of the material, technical conditions of the network—the wide range of output devices for printing and display that were getting hooked up. It was as if the number and metrical length of lines in the sonnet had been suggested by the diverse practices of Renaissance papermaking.

And if the heterogeneity of output devices for printing and display had certain ramifications for online, networked documents, so did the heterogeneity of input devices. In November 1970, when the ARPANET had grown to more than ten nodes, the NWG met in Houston at the Joint Computer Conference to discuss its use and further development. Participants worried that the ARPANET did not yet have "significant operational usage," and were agreed that the availability of documentation would offer a crucial encouragement to potential users. According to RFC 77 (November 20, 1970), they decided to concentrate for the moment on keyboard input devices with printouts rather than try to tackle both keyboards and graphic devices like light pens and mice with screen display output. But they also worried about the prestige of keyboards:

[Doug] Engelbart: Perhaps we should put off graphics several months so as not to delay typewriters. Typewriters are important.

[Edwin Meyer]: But would that be sufficiently impressive for DOD [Department of Defense] people?

Even limiting the discussion to keyboards meant that there were some thorny issues yet to resolve. Engelbart of SRI reported that the NIC was setting up a "dialogue system" that included an online documentation service, and might even put "a communication agent and technical liaison officer" at every node. One thing he needed to know in order to get his documentation service off the ground was what sorts of "console interaction" each site had.[39]

Keyboard-based console interaction came in four basic types. It could be either full or half duplex, and either character or line oriented. A full-duplex system can transmit and receive at the same time (like a telephone), while a half-duplex system allows transmission in one direction at a time (like a walkie-talkie). Character-oriented systems respond to characters as they are typed, while line-oriented systems respond to complete lines of type that end with an end-of-line command, like a carriage return. The system response in question is called echoing, or as Jon Postel understood it from Engelbart's discussion, "the situation where the console, a peripheral processor or some very low level software, echo[e]s the character [or line that is] entered." Different nodes on the network handled input in different ways; some relied on echoing at a local level, while others echoed remote input themselves. The difference had less to do with what one communicated to the node than it did with how one communicated. At what level did the character or line entered actually enter the system in question? Precisely how and *where* was electronic text bibliographic or embodied? The implications of differences among systems consumed much of the conversation during the second day of the Houston meetings, where "it was discovered" that many participants had no idea "where their own systems fit" into this scheme of things. NWG members were "very vague about what it means to interact with a system . . . different than their own," and some felt "threatened by the revelation of their documentation inadequacy."[40] Different nodes were lagging in their documentation of specifications integral to documentation itself.

Just as the diversity among output devices had eventual implications for the textual standards and practices of the ARPANET, this initial inkling of diversity among input devices on the part of Postel and his peers helped to reveal where standards and protocols remained uncertain. Tacit experiences of text *as* text, of lines and characters entered, had to be revised, broadening them to acknowledge one's "own" means of entering them in distinction to somebody else's different means of entering the same thing. Some distinctions were obvious, like the kind and model of keyboard device being used (ASR–33 Teletype machines were for a time the most common).[41] Many much less evident levels of distinction remained necessary to document, however, and these accordingly eroded some of the previously and naively self-evident qualities of online text as text. Licklider had recognized the very same issue in much broader terms when he chafed at the implications of the word *interface:* "'Interface,' with its connotation of a mere surface, a plane of separation between the man and the machine, focuses attention on the caps of the typewriter keys, . . . but not on the human operator's repertory of skilled reactions and not on the input-output programs of the computer. The crucial regions for research and development seem to lie on both sides of the literal interface." Licklider favored the term *intermedium* to refer to the whole depth and density of human-computer interactions that would have to occur at the levels of hardware, software, and users' "organs and skills." "Once we assume that definition of the domain," Licklider (1990, 92) realized, "it is impossible to draw a sharp line between [its] nonlinguistic and linguistic parts." Thanks to Engelbart's question about console interaction, the NWG had the sharpness of the line between text and technology thrown into question.

Two years later, the blurriness of the line had become normalized, part and party to the practices of networking on the ARPANET, which had by then grown to almost thirty nodes.[42] Still concerned that the network was not being used to its full potential, ARPA arranged a public demonstration at the International Conference on Computer Communication held in Washington, DC, in October 1972. Members of the NWG made the arrangements. There were demonstration sessions and film screenings to introduce the network as well as a roomful of terminals—one IMP-like computer (called a TIP) with more than forty different input/output devices attached—where conference goers could try out the ARPANET themselves.[43] In order to facilitate use by outsiders, the NWG supplied "parameter sheets" at every console so users could set their terminal appropriately along with a booklet titled "Scenarios for Using the ARPANET."[44] Each of the nineteen scenarios begins with a series of commands that users had to enter in order to configure the terminal "to suit the serving HOST"—that is, to communicate effectively with the distant mainframe computer involved in each scenario. These instructions explained whether the host "prefers to do its own echoing, a character at a time," or "interacts line-at-a-time" and "assumes local echoing" by the terminal in Washington. Users also needed to know whether the host could "distinguish between upper and lower case alphabetics," and what special characters the host used as system prompts and command keys. The practices of networking now involved much more intermedium and less interface, as the variables of terminal settings became routine.

The scenarios were offered in a tone similar to the one that characterized the early RFCs. In his opening, Bob Metcalfe acknowledged that every effort had been made to "make them accurate," but he was "certain that they still contain errors." He assumed that conference goers would attend demonstrations, and urged them to "approach the ARPANET aggressively yourself" and ask for help when needed, "rather than stew in your own juices." The aim was to introduce participants to some of the "possibilities in computer communication" at this early stage in its development. They were prompted "to browse" the network, sampling its variety. The scenarios were listed "by (approximate) category," and ranged from games and "conversational programs" to database queries as well as remote programming and processing with different computer languages and applications.

The informality of "browsing" the network—a phrasing that is still so current—ran counter to many of the prevailing assumptions about what the ARPANET was for in 1972.[45] Early and novice users might have entertained themselves playing LIFE or CHESS, but the main point was clearly to connect and share the extraordinary investments of capital and labor that had been made at individual sites on the network. From the conference hotel in Washington, users could log on to mainframe computers at Harvard, MIT, BBN, the University of California, Los Angeles and Santa Barbara, the University of Utah, SRI, and the Stanford Artificial Intelligence Laboratory (SAIL). These were machines worth hundreds of thousands of dollars at which teams of engineers and other specialists had worked for years. The scenarios demonstrated that such resources—hardware, system architecture, and programming—could be shared remotely and on the fly, in real time. Some of the online applications offered as samples to prove the point were already well-known within computing circles, like Joseph Weizenbaum's ELIZA at Harvard, a conversational program in which the computer emulates a Rogerian psychotherapist. (SAIL had a comparable program in which the computer emulates a paranoid mental patient.) Others were new; Vint Cerf remembers, "You could get into the databases [at SRI] and get documents, RFCs, and things like that."[46] There were rudimentary ways to send "mail"; there were ways to write, assemble, load, and run programs in FORTRAN; there was an example of symbolic algebraic manipulation; and the SAIL computer could make a live connection to the Associated Press wire service and then process the reports it received into a searchable archive. The sample search described was for the keyword "nixon"; the Nixon–McGovern election was two weeks away.

In hindsight, it has been all too easy to note that the ARPANET community had effectively misperceived the purposes of digital networking. Like the hoots and declamations that characterized demonstrations of Edison's tinfoil phonograph in 1878, many of the scenarios of 1972 seem off the mark, but only if one knows what happened next. Intensive resource sharing of the sort promoted at the International Conference on Computer Communication never developed in the manner expected. Only with the extensive adoption of e-mail would the network demonstrate anything like its eventual success as a communication medium, and even then its use remained haunted for a time by other, lingering associations. Using the ARPANET felt "rather like taking a tank for a joyride," recalls Janet Abbate (1999, 2), or "a bit like being a stowaway on an aircraft carrier," according to Katie Hafner and Matthew Lyon (1996, 188). The informality of e-mail and browsing the Net would become the norm, but it would apparently take a few years for their countervailing military-industrial tenor to be forgotten or reengineered, as internetworking added to the scale and scope of the original network, newer and newer new users continued to get online, and accordingly, a new media public began to emerge.[47]

The "Scenarios" booklet is itself a curious document. It was eventually issued as RFC 254 (October 29, 1972) and also identified as NIC #11863, although it appears never to have been digital or digitized, or to have been republished.[48] The booklet contains scenarios in the dual sense that it describes imagined interactions between users and computers as well as in the more specific sense in which motion picture screenplays used to be referred to as scenarios. Like a shooting script, that is, these scenarios consist of elaborate directions for performing the interactions they also partially, lexically, and typographically represent. What this means in the language of computer text-encoding today is that the scenarios are self-consciously "marked up." They are full of typographic cues that distinguish different kinds of content: any character that a user is expected to type—including punctuation, but not including spaces or command keys—is "underlined to set it off from computer type-out, general instructions," and comments that appear in italics. Spaces and carriage returns that a user is expected to type appear as [SP] and [CR] in enclosed boxes that make each look like the top of an individual typewriter key. Spaces that a user is not expected to type also appear on the page, but only for "readability," as if to acknowledge that off- and online interests in textual representation necessarily diverge. The paper scenarios (in the narrow sense) describe performed scenarios (in a broader sense), which got recorded onto a paper printout as they were performed. The printout itself could contain only a limited range of typographic cues, usually letters, numerals, spaces, and a small number of other characters, like punctuation marks, "@," and "*." Like *Libraries of the Future,* the "Scenarios" booklet relies on typographic information—underlining, italics, and special characters—that the digital system it represents could not.

The "Scenarios" booklet refers explicitly to documents, but it does not actually contain, quote, or reproduce any of them. The fifth scenario consists in part of a "scenario for [the] NIC document locator and browsing system," which allowed users to retrieve "selected documents online," including RFCs. It was an application "normally used by people with some knowledge and experience in using" the Online System (NLS) at SRI, but was nonetheless offered to conference goers, who could also attend a more formal demonstration session about the NLS and its capabilities. The locator and browsing scenario lists the steps required to enter the NLS, access the locator application, list documents, examine the table of contents for a selected document, and "load and print a particular file" "containing" the desired part of a document. Users are given the keystrokes to enter at every step, but—in this scenario—no exemplary system response. To the extent that documents would have appeared within the enacted scenario, they would have arrived on paper or screen, character by character, and nested within the prompts, menus, commands, and file names that also would have been printed by the terminal. If the "Scenarios" booklet resembles a shooting script, then, this scenario contains the lines for only one role. But the printout or display produced would have been both dialogic and miscellaneous—a list, a table of contents, part of a document, and its file name (for example, "<nic>LOC7440.nls;8"), punctuated repeatedly by prompts (in this case "*") and commands like "pb.2[CR]xbm[CR]." This is a partial recreation of the scenario in question. The left-hand column is from the published scenario (like a shooting script), while the right-hand column represents the enacted interactions of user, host, and terminal (like the resulting movie) as they would have appeared serially on paper or on screen.

```
@nls [CR]
*load file<nic>locator [CR]
*print branch .2 [CR]
*print branch .STATEMENTNUMBER [CR]
xeb [CR]
*print branch .STATEMENTNUMBER [SP] ↑ [CR]
```

```
@nls
*lf<nic>locator
*pb.2
xbm
*pb.STATEMENTNUMBER
xeb
*pb.STATEMENTNUMBER↑
[The document requested
would print out here]
<nic>LOC7440.nls;8
*
```

The online document would be there, on a computer in California and via the TIP in Washington, but to read it as it appeared would involve parsing its typographic reproduction from the typographic context of its reproducing.[49]

Electronic documents in 1972, like those in Licklider's 1965 prognostications, appeared amid and are therefore distinct from commands, prompts, headers, menus, and messages. They were composed of typed lines and characters, they could be transmitted as packets, and they were contained in files. More important, what distinguished them most as documents was neither an essential, ontological property nor a material, bibliographic difference, but rather their social or cultural standing. In Licklider's terms, they were "items of basic interest," whether they were RFCs or print publications converted into electronic form. They were identifiable as documents, that is, because of the relative importance—the meaning—ascribed to them by the limited social network within which they circulated (or potentially circulated). Their importance was "relative" both in their distinction from other online texts—commands, prompts, headers, menus, and messages—as well as in their comparison with other electronic documents, their location within document series, and their association with the necessary electronic catalog tags or descriptors. But their relative importance was also derived outside the system, and according to the ways in which the limited social network of their circulation both mirrored and modified the values attending the dominant cultural economy of the late 1960s in the United States.[50] Documents were espied by users amid the typographic clutter of commands and prompts in part because their status as documents depended a priori on their use or usefulness, where "usefulness," like "importance," of necessity implied the normative interests of cold war R & D and the emerging field of computer science, and thus must also have implied various and countervailing cybernetic anxieties and campus tumults.

The question of what electronic text fundamentally *is* had come up with force, and it would keep coming up in myriad unacknowledged ways as users helped to define the new medium as, in part, a means of textual communication as well as a tool for text formatting and display. Documents, as a particular class of electronic text, were still few and "plain vanilla," as ASCII files have come to be known in Internet slang. In addition to RFCs and other network/ed documentation, in 1971 a precocious undergraduate had typed the Declaration of Independence into a University of Illinois computer, all in capital letters, eventually making it available to everyone on the ARPANET. That was the beginning of Michael Hart's Project Gutenberg (still publishing today). Thus, even before the 1972 scenarios had their debut, Hart's scenario for the network suggested that it be used to disseminate as many documents as users of the network might desire. A document was just what Hart thought important to enter as data. In retrospect, it is hard to tell to what degree or in what combination his scenario was subversive—against the rules or just counter to the normal use of the mainframe, for instance, or declaring independence from the Pentagon's cold war logic amid controversy over its influence on the Champaign-Urbana campus.[51] It is likewise hard to tell whether or in what sense his Thomas Jefferson scenario arose instinctively from the same impulse—a similar association of new media and cultural authority—that had driven Edison to suggest "our Washingtons" and "our Lincolns" as subjects for the first phonographs so many years before.

Graphical Barbed Wire

I have been suggesting in this chapter that a cluster of different events in the late 1960s and early 1970s in the United States might be thought of together as experiences of bibliographic fact. The events may have clustered by coincidence, but inasmuch as bibliography was significant to each, these experiences shared common elements and implications. True, *bibliography* is a grandiose term for what may seem like a tenuous thread connecting an arrest in Boston, a student protest in Berkeley, a television studio in New York, a report on libraries, and a series of memos about the origins of packet-switched networking. The word *bibliography* is more commonly used to refer to a seemingly old-fashioned pursuit of academics and antiquarians, something specialists do when they describe books and list imprints. But one of the things I want to note in conclusion is that traditional bibliography—in that old-fashioned sense—did play a role in both the self-imaginings of U.S. culture during these years and, more broadly, the contexts of mediation. The qualities of books as physical bodies formed a subject of public debate and even controversy, and as such they formed one of the cultural sites from within which digital documents and networks would emerge into and partially constitutive of a changeable new media public.

The controversy began when Mumford (1968) reviewed six volumes of *The Journals and Miscellaneous Notebooks of Ralph Waldo Emerson* for the *New York Review of Books.* In his review, Mumford excoriates the editors for their neglect of readers and, by extension, Emerson, at the same time that he grudgingly acknowledges their skills in descriptive bibliography. The review is titled "Emerson behind Barbed Wire," and the barbs in question are an assortment of "twenty different diacritical marks" (4) that the editors had used to help them represent in print the precise linguistic and bibliographic information contained in "all of the surviving documents" by Emerson. Emerson's handwriting is transcribed, and there are marks to indicate erasures, cancellations, and insertions in the manuscripts as well as damaged pages and variant versions. The result is an exhaustive edition characterized by its own "ruthless typographic mutilation." Mumford quotes an exemplary sentence by Emerson: "The best visions of the Christian <are> ↑ correspond ↓ cold ↑ ly ↓ & imperfect ↑ ly ↓ to the promise of infinite reward<s> which the scripture | contains | reveals |." What is lost by such scholarly editorial practice, according to Mumford, is readerly access to the great American author:

> Oh! But Emerson is there! One sees his figure at a distance, through a barbed wire entanglement of diacritical marks; the searchlight from the control tower, meant to keep Emerson from escaping, or even making a movement without being noticed by the guards, keeps on sweeping into the reader's eyes and blinding him; the voice in which Emerson faintly calls out to one is drowned by the whirring of the critical helicopter, hovering overhead, . . . Yes: Emerson is there. But after an hour or two of trying to find an unguarded place in the scholarly enclosure where one may get near enough to him for a little uninterrupted conversation, one gives up in despair, and departs, as one might from a futile visit to a friend in a concentration camp.

Mumford's prose may be purple, but his point is clear. There's no getting at Emerson behind all of that bibliographic barbed wire. An edition "*by* scholars and strictly *for* scholars" flies in the face of "literary values" as well as "humanistic aims."

One curious thing about Mumford's critique is his appreciation of exacting bibliographic description. He may hate the result, but he does acknowledge that—for all of its "typographic mutilation"—this edition ends up "as near as print can get to a photographic copy of the original" manuscripts. Mutilation *and* photographic copy? How can that be? The apparent contradiction arises in part from Mumford's unstated distinction between work and text, his assumption that material texts—whether the published transcription or the original manuscript pages—are merely impoverished representations of Emerson's intangible mental work. Representations are only a technological necessity; it is the author's work, his thought, that readers must engage. And in Mumford's ken, readers do so in metaphorically vocal terms—"a little uninterrupted conversation"—no less than Moses Coit Tyler's American literary history of ninety years before depended on the metaphoric figures of printed speech and literary lispings. Conversation with genius was still the point, and no "technological extravagance," no ultimate typographic scheme, in the production of an edition could really help with that.

Mumford's review provoked much comment in the Letters column of the *New York Review of Books.* Then Edmund Wilson weighed in with a two-part article in the same periodical titled "The Fruits of the MLA [Modern Language Association]" (1968). Wilson went much further than Mumford. He launched an attack on the MLA for the editorial practices of its Center for Editions of American Authors (CEAA), which was then engaged in the publication of critical and documentary editions along the lines of the six-volume Emerson that had so piqued Mumford. Wilson condemned the CEAA as a "boondoggle," a waste of the taxpayer money granted it by the brand-new National Endowment for the Humanities.[52] (Wilson himself had long favored a less scholarly publication endeavor, based on the French Pléiade editions.)[53] More bitter letters followed, with extensive replies and vituperative rebuttals, so that the ensuing controversy was described amusedly as "a full scale litry *[sic]* rumble" in the pages of the *New York Times* (1971), and eventually an instance of regrettable "carnage" in the *Papers of the Bibliographic Society of America* (1974).[54]

Many of the niceties of this brouhaha are not relevant here, but some of the larger questions the controversy raised do resonate powerfully with the microhistory of the ARPANET sketched above. One is certainly the relationship between the academy and U.S. public life. How and by what measure are the labors of university researchers effective in and effected by their larger social contexts? Another related issue is that of federal support for academic research. What are its ideological components? Are they necessary? And how should scholarly outcomes be assessed in the short and long term? But more specifically, the "Fruits of the MLA" controversy, as it came to be called, raised questions of bibliographic facticity. Indeed, whether one picked up Emerson's journals in their six-volume incarnation of 1968 or picked out an RFC online in 1972, what one faced was a new material text, a new bibliographic body, insisting that its reproduction of documents was representational. The arrows, brackets, and barbed wire of the edition, and the prompts, commands, and barbed wire of the printout (or on-screen) RFC were both specialized codes or jargons, although *jargon* is a bit of a misnomer in the sense that barbed wire is almost never pronounced, never uttered.[55] It exists by definition on the page (or on the screen) graphically to represent—to mark up—the bibliographic qualities of Emerson's manuscripts or RFCs.

The editors' barbed wire represents rips, erasures, and emendations, the varied material traces that inform Emerson's original manuscripts, and thus help to provide or impede access to Emerson's thought. Meanwhile, the network users' barbed wire represents file names, application access points, and host and terminal interactions, the varied textual markers that literally ("operationally") made digital documents appear as such on printouts or screens. This is not to suggest any exact equivalence between the invention of the ARPANET and the fortunes of the National Endowment for the Humanities or the MLA, only to point out some of the common soil from which they each bore different fruit, and at the same time to underscore that traditional questions of bibliography are particularly germane to the task of writing any history of the ARPANET and its successor, the Internet.[56]

Today Internet documentation is available online via the RFC Editor (<www.rfc-editor.org>), a publication funded by the Internet Society (<www.isoc.org>). "At some point in [their] history," however, the first six hundred RFCs were "lost." Many that had existed electronically, whether as first- or second-generation electronic objects, were not ported forward as the relevant hardware and software changed. Shortly before his death in 1998, Postal initiated the RFC-Online Project, to republish the missing RFCs by typing or scanning from extant hard copies. At first, the staff and volunteers of the RFC-Online Project simply tried to make their republished versions "look as much like the original as possible," with the exception that ASCII diagrams replaced hand-drawn figures and single spacing replaced double spacing. But "looking like" is hardly a simple matter on the network, as the authors of the first RFCs themselves so early became aware. Different output specifications—different monitors—make a difference, as do different network protocols, software choices, and software versions, now including different Web browsers and browser updates. Lately, the RFC-Online Project has shifted its objective. The project has stopped aiming for an old look, and instead follows the current RFC format rules "while preserving the contents as strictly as possible." The scanned or retyped RFCs are handled just like new ones: processed and edited by the editorial staff of the RFC Editor. Specifically, the files are "nroffed," processed using an adaptation of the UNIX "nroff" program that only became available after many of the early RFCs in question first appeared.[57] Some contain bracketed notations indicating partial provenance—for example, RFC 4, "[This RFC was put into machine readable form for entry][into the online RFC archives by David Capshaw 11/97]"—but some, like RFC 3, do not.

Can these early RFCs, now digitized or redigitized, really be considered identical with the original RFCs, as I have admittedly regarded them in writing this chapter? In what sense is the RFC 3 I read today equal to or identical with the RFC 3 of 1969, if I have found the former as an nroffed .txt file via my Web browser on today's substantially different and vastly more ample Internet? For one, I read it now on a networked personal computer—a reading process and context that are both influenced by today's "what you see is what you get" (WYSIWYG) paradigm.[58] Readers, software, and hardware now habitually handle electronic text in much different ways than they did in 1969. Barbed wire is usually obscured, unless I opt to "view source" or "reveal codes" in some way. My screen is a bitmap; my letterforms—not just my letters—are generally described in code. And no physical remnant of the old ARPANET still exists to store electronic files or route packets; today, the work of the old IMP happens somewhere inside my own personal computer. Each of these differences, and many more like them, would require explanation in order to unfold a complete bibliography or narrative of provenance for the RFC 3 I read today via the World Wide Web. That is a responsibility that I have quietly shirked in treating the old and new RFC 3 as the same thing, the same body. In effect, my endnote citations have collapsed 1969 into today. The following chapter explores this and similar species of temporal collapses, wormholes, and dislocations, considering the ways in which the World Wide Web today functions as a new medium for history, a potential new instrument of historicity.

4 New Media </Body>

The Internet of 1854

In 1998, historian Roy Rosenzweig observed that the *New York Times* had mentioned the Internet only once before 1988. Though his finding might have been based on reading the *Times* and its index, or a certain amount of cranking through microfilm, Rosenzweig used a Lexis/Nexus database to search the *Times*.[1] Searching today can be yet a different matter, since the ProQuest Company has made its full-text version available to customers on the World Wide Web. ProQuest scanned the newspaper from microfilm and linked the resulting digital images to their linguistic content in an underlying ASCII text, which can be searched "through the ProQuest interface."[2] So users who belong to a ProQuest subscriber group can now search the *New York Times* as an online database and retrieve PDF images of articles and pages from 1851 to 2001. There are wrinkles, however. A quick search for "the Internet" reveals seventy-five mentions before 1988, the earliest in an advertisement for patent medicine published in September 1854. The scanning technology used by ProQuest has misread "the interest" on microfilm as "the Internet," here and on numerous other occasions. So the *Times* did not mention the Internet in 1854, except insofar as ProQuest—that is, the Internet—is sure that it did. Neither the microform version of the 1854 advertisement nor its digitized image include "the Internet"; these words appear only in a remote ASCII transcription, searched but unseen, as well as in the query box where a researcher has typed them into her Web browser.

Like a Freudian slip, the Internet of 1854 is a fleeting, if symptomatic, incoherence. Researchers stumble and move on, as if over the incidental irruption of an active unconscious. But what psychic debility, what repressed urge, can explain this errant search result? The Internet of 1854 illustrates the limitations that presently attend optical character recognition (OCR). The scanner chronically "misreads," not because of any hardware malfunction or programming error but precisely because scanning is not reading. If the Internet of 1854 expresses an unconscious desire, then, it is the long-lived anthropomorphic dream of a reading machine and a self-apprehending text. And it is this dream that helps to deflect researchers' attention from the real human agents involved, like the typesetters and printers of 1854, the microfilm camera operators and film processors of the twentieth century, and the scanner technicians and data handlers employed today by ProQuest's offshore contractor. Add to these a talented cast of what Bruno Latour would call "nonhuman participants"—like metal sorts of type, I, n, t, e, r, e, s, t, all variously worn and inked; a paper copy, possibly creased or stained; a frame of microfilm, possibly scratched or overexposed; as well as several generations of saved electronic data, "dirty" and partly cleaned—and this fleeting parapraxis begins to speak volumes.[3] Like an errata sheet bound into an early modern book, today's errant search result reminds users that the complete work, the *New York Times* in this case, is less of an "autonomous object" than the ongoing result of its own making, remaking, and reception.[4]

My point is only incidentally that the Internet is wrong about its own history. More important, ProQuest's "Historical Newspapers" and Web-based documents like them are "historical" in some interesting ways. In one sense, the pages of the *New York Times* that appear in desktop windows are what Vivian Sobchack (2004, 306) might call "distilled images": they are intensely labored points of contact between the present and the past, a collision and an overlap of different times and formats.[5] In another sense, however, these images are diluted not distilled: they are scans of films of pages, attenuated in their stages of removal from a unique original printed in 1854. Yet to the reader whose research interest happens to coincide with "the interest" of 1854, in this case, neither distillation nor dilution needs to be acknowledged. ProQuest's *Times,* like the microfilm before it, gets cited simply as the *Times.* The researcher's footnote assumes as well as performs an identity between the original of 1854 and its later reproductions, as if a paper copy were equal to and identical with its image on a screen.

Nor are the same concerns moot when the Web-based documents in question are first-generation examples, created for the World Wide Web itself, rather than laboriously imported from older media. For example, the Web site of the World Wide Web Consortium (W3C) offers "A Little History of the World Wide Web." The little history identifies both "the **first web page**" and "the least recently modified web page," both from 1990 and CERN, the European research center for nuclear physics where Berners-Lee invented and named the World Wide Web.[6] The first Web page is identified as <http://nxoc01.cern.ch/hypertext/WWW/TheProject.html> without being quoted

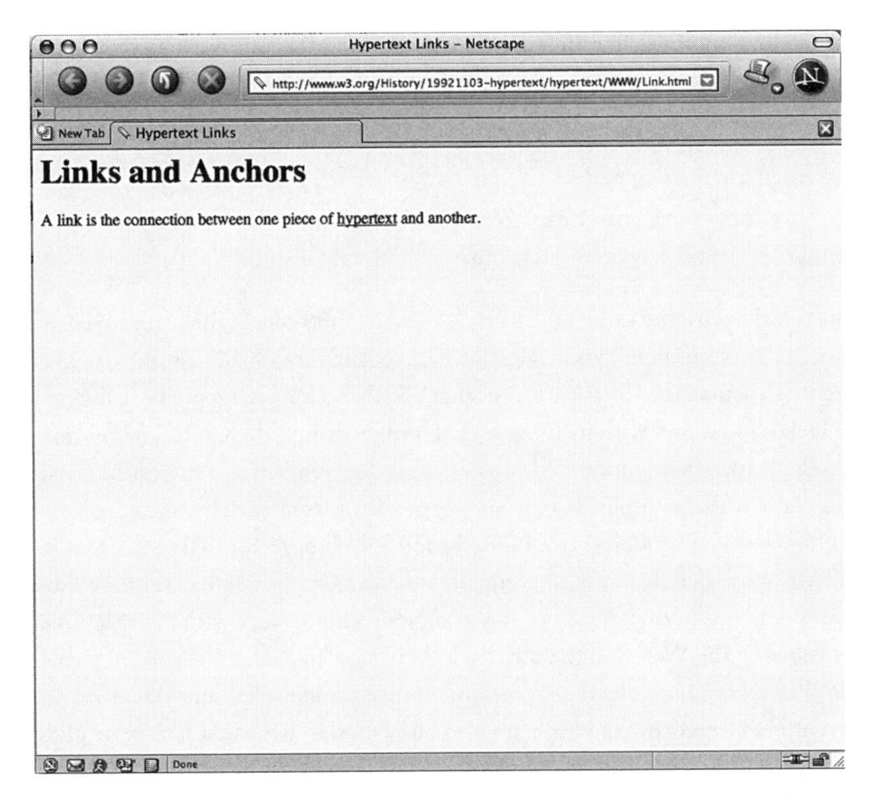

Figure 4.1 The least recently modified Web page, 1990–2005. (*Source:* <www.w3c.org>.)

or described, because, unfortunately, "CERN no longer supports the historical site." The least recently modified web page does still exist, however, and can be viewed by users of the Web today. The underlined words represent a hyperlink, which leads users to another page on the W3C Web site. The least recently modified page turns out to be a twelve-word definition of hyperlinks: "A link is the connection between one piece of hypertext and another."

What is "least recently modified" about this page? How—in other words—is it historical? Its location has changed, since it now resides on a server at W3C and not at CERN. Its unique identifier—its name, if not its title—has changed, since a new location requires a new URL. Its context has also changed, since the words "least recently modified" could not possibly have signified the hyperlink leading users to it in 1990. And its appearance has changed, since it now opens via a Web browser that did not exist in

1990 and, likely, on a monitor that didn't either. One thing that has not changed is the underlying HTML source,

```
<title>Hypertext Links</title>
<h1>Links and Anchors</h1>
A link is the connection between one piece of
<a href=WhatIs.html>hypertext</a> and another
```

The words and markup are the same now as they were in 1990, when they were written and saved on a CERN computer. This makes the least recently modified page historical in quite a different sense than the "historical site" that CERN no longer supports or the first Web page, which is identified with its unique URL rather than as the words and markup it contained. The least recently modified page is offered to readers as a historical document within a context that complicates the very grounds of its historicity.

A column of newsprint rendered as a PDF file and a Web page written in the first iteration of HTML are very different electronic documents. One is a digitized image, and the other a first-generation digital object, born digital. One is data within a relational database accessible via the Web, and the other is a Web page linked among related pages. Despite their differences, both electronic documents are presented on the Web today as evidence, as matters of record. And both are also presented in ways that ignore or elide the physical—that is, the bibliographic—properties of documents that I argued in chapter 3 were so at issue in U.S. culture just a few decades ago. What has happened? One might say that users of the World Wide Web today resemble the students in an art history lecture who sit facing a screen and consuming slides of prints of paintings as if they were the actual paintings themselves. Students know they are not seeing real paintings, of course, but the habitual contexts of display suggest otherwise.[7] Or one might say that users of the World Wide Web today resemble the characters in a Don DeLillo novel (1985) who are visiting and taking pictures of a tourist attraction known as "The Most Photographed Barn in America," since the attraction becomes increasingly reified, more and more "most photographed"—consider, "least recently modified"—and less and less crucially a physical barn. These analogies aren't perfect but do help to portray how, as David Weinberger (2002, 35) puts it with such gleeful imprecision, "Web documents are weird."

This chapter seeks to describe and contextualize the weirdness of documents on the World Wide Web. In particular, it asks how electronic documents work as evidence, and how the World Wide Web both is and is not a temporal medium. These questions have profound implications for the ongoing definition of publication, and thus of contemporary media publics and public memory. My ungainly chapter title, "New Media </Body>," uses an HTML tag to revise the title of chapter 3 ("New Media Bodies"), and in doing so underscores that bibliographic facticity seems curiously mystified in the digital environment. Web pages today regularly have explicit bodies, content, and markup wedged between two tags, <body> and </body>. And according to the W3C, the whole World Wide Web itself has "a body of software."[8] But these bodies are not easily discerned as or in relation to either bibliographic fact or historical evidence. The body of a Web page stands in distinction from its header, regularly indicated by the tags <head> and </head>, as well as from a line of HTML that self-describes the HTML version information or document-type definition required for browsers to display the page. These "bodies" are not actual texts, then; they are only one part of one kind of text available to users of the Internet, where the ontology of text itself (What *is* electronic text, after all?) remains open to question, even—or especially—among scholars for whom textual criticism and textual editing are specialties.[9]

The subject of historical evidence has lurked in the background throughout the previous chapters, the last of which describes a particular and particularly broad context for the origins of distributed digital networks in the United States. The current chapter pulls questions of evidence to the foreground in order to return to the issues of reflexivity addressed briefly in the introduction above. How are media the subjects of history when doing history depends on so many tacit conditions of mediation? More specifically, how might present attempts to historicize the Web be complicated by the uses and characteristics of the Web itself? How does or how can the Web work as evidence of its own past? While this chapter calls on the introduction, it also stands as a sequel to the previous chapter, in which the "Scenarios for Using the ARPANET" suggest so much about the social, intellectual, and political contexts of 1968–1972, but so little about today's cyberreality. Like chapter 2 ("New Media Users") in relation to chapter 1 ("New Media Publics"), the present chapter pursues the emergent social meanings of a new medium among its uses and users. Chapter 2 follows the new medium of recorded sound into the varied cultural economy within which it became defined primarily as a mechanical amusement for the repetition of prerecorded musical records. The present chapter follows the new medium of digital networks into the varied cultural economy within which the global Internet is commonsensical according to the continuing variety and intensity of its uses, among them the creation, transmission, and display of electronic documents, if also the contested circulation of musical files, the buzz of e-commerce, the instantiation of virtual communities and social networks, and all of the other forms of "knowledge work" that the Internet has so recently come to demand.[10]

Let me be clear about my vocabulary. I am using the term Web *page* to refer to files written in markup language that have URLs and can be displayed on screen by a web browser in keeping with the client/server architecture of the World Wide Web. "The World Wide Web of the 1990s foregrounded the page as a basic unit of data organization," notes Lev Manovich (2001, 16), even though these new pages commonly contain many elements beyond those of the traditional codex page, including sound and moving images. Recently it has also become common for Web pages to contain dynamic content—that is, content unique to a session and a user because generated by a gateway application and database that are used to produce a menu of search results, like a list of images from the *Times* or a list of auctions on eBay, or used to produce customized pages, like those that greet repeat customers at Amazon and follow them around with a shopping cart. A Web page, to generalize, is thus partly a matter of format and partly a matter of delivery. According to this definition, the World Wide Web contains or points to many elements that are not specifically Web pages, and many Web pages contain elements—sounds, images, applets, and applications—that are not written in markup language but are named within a page that is. There are private pages and public ones; there are searchable, surface pages and "deep" pages that cannot yet be crawled and therefore searched by Google or its competitors.

In contrast to the word *page,* I am using the term electronic *document* without reference to format. As the observations of the previous chapter make clear, an electronic document cannot be identified with any exclusive material property, any bibliographic difference that distinguishes it from other electronic objects; it can only be identified according to its cultural standing, its meaning, within the social network of its potential circulation. An electronic document is any electronic object that is used to document, that stands as a potential ally in explanation. So while the Web page is "a basic unit of data organization," an electronic document is Licklider's "item of basic interest." One is an issue of format, and the other is an issue of context, a matter of concern.[11] So some Web pages work as electronic documents (for example, the "least recently modified" page above), while some pages more importantly point to documents (ProQuest points to images of the *Times*). The World Wide Web itself might even be called "the largest document ever written," if it is called that within a context where its own meaning as such is at stake, as it is today for archivists concerned with preserving the Web for future researchers.[12]

When I use the word *today,* I mean "as of this writing" (2005), and I am cognizant of the humility that requires, since readers—as of their reading—will know so much more. By *history,* I mean to invoke the limited sense in which doing history means putting together narratives about events (also "history") based on interpreting the "indexical survivals" or inscriptions that form a fragmentary record of the past. Put this way, history is not a science but a hermeneutic, an interpretive mode sutured to a disciplinary practice.[13] It is the province of the humanities proper, the scholarly fields, cultural institutions, and social practices that attend—construct, perpetuate—and interpret the past. History as such has had its own modern history coincident with and somehow codetermined by modern inscriptive media, particularly cinema in Philip Rosen's (1994, xix, 143) recent account, and particularly photography in Walter Benjamin's.[14] And—no surprise—history has lately been mourned by critics on both the Left and Right, as another victim of postmodernism. Cellular phones, news crawls, and channel and Web surfing help to promote and transfigure the synchronic, inspiring what Fredric Jameson (2003, 707) calls a "new nonchronological and nontemporal pattern." Everything happens all at once. But digital media are not just agents of postmodernism in such accounts, they are equally its expressions: "What could not be mapped cognitively in the world of modernism now slowly brightens into the very circuits of the new transnational cybernetic" (701). This is a particularly bleak view that underscores the present importance of questions like the ones framed here. Is the history of media—or indeed, the history of *anything*—possible amid the synchronous postmodern glare?

Just as critical theory thus maps its own course against the history of media and the meaning of *history,* related tussles over the same term have abounded in public culture more broadly, during the so-called culture wars of the 1990s in particular. By this light, the World Wide Web emerged to popular consciousness—the Mosaic browser in 1993, and the Netscape IPO in 1995—at the same time that vitriolic disputes erupted over the National History Standards Project (1994), for instance, as well as the Smithsonian Institution's Enola Gay exhibit (1994–1995) and Disney's plans for a theme park near Civil War battlefields in Virginia (1994). Less controversial but just as telling, *history* simultaneously signified and continues to signify a competitive array of consumer goods, the bourgeois uses of culture enabled and in a sense "overproduced" by heritage sites as well as television, video, and cinema, by the History Channel (1995–), for instance, as well as variously "historical" films like Oliver Stone's *JFK* (1991), Steven Spielberg's *Schindler's List* (1993), and Robert Zemeckis's *Forrest Gump* (1994).[15] In the heightened rhetoric of the 1990s, the Web was either a "brave new world or [a] blind alley." The most troubling feature of the Internet for the culture warriors was the pervasive lack of distinction made on the network between "the true and the false, the important and the trivial, the enduring and the ephemeral."[16] Many, on the Left and Right, had just the same criticisms of *JFK, Schindler,* and *Gump.*[17]

Lastly, by *the Internet* and *the World Wide Web,* I mean both the architecture of each network as well as the massive diversity of material they differently connect and contain. (Internet architecture includes both hardware and software, while the architecture of the World Wide Web consists of software that works on and via the hardware of the Internet.) Size and complexity both call for much more humility, since no single chapter, no single book, can really address or analyze the whole Web. In fact, analyzing the World Wide Web curiously resembles making a Web page. This is not to resuscitate "obsolete" accounts of readers supposedly empowered by hypertext because they can choose which links to follow.[18] Rather, it is to notice the obvious: every analysis of the World Wide Web—including the present one—selects examples, quotes excerpts, and assembles links. Making a Web page means—in part—doing just the same sorts of things: selecting and pasting in content as well as assembling links. The mirror effect should cause anxiety. Like an assiduous anthologist, the Web critic's presumptive "ambition to represent a whole through its parts is always undermined by readers' awareness that the parts have been chosen for their difference from those left out" (Price 2000, 6).[19] This is one reason why Jeffrey Sconce (2003, 191) is so right to question what he dubs "the more vapory wing of new media studies" for its rapt attention to digital artworks or examples of Net art "as evidence of significant transformations in culture and society." Selecting singular examples from the World Wide Web in order to support claims about the Web or digital culture as a whole is a lot like manufacturing one's own evidence, minting one's own coin. Of course, criticism of all kinds depends on the critic's discretion, on a reasoned and reasonable balance struck between the representative and the anomalous. But the sheer size and diversity of the World Wide Web suggests that in this case, balance might be a rank impossibility.

One hedge against the "cult of the anomaly" (Price 2000, 6) or runaway "anomaly fatigue" (Weinberger 2002, 15) is to take a longer view, to focus on tools, methods, and protocols rather than the dubious exemplarity of Web pages themselves. Another strategy is to turn anomaly against itself, to concentrate on error or errant results, like the Internet of 1854, which can reveal the assumptions that lie behind different uses of the Web. As Carlo Ginzburg (2004, 556) puts it, no stipulated "norm can predict the full range of its transgressions; transgressions and anomalies, on the contrary, always imply the norm and therefore urge us to take it into account as well. This is why a research strategy based on blurred edges, mistakes, and anomalies seems" so rewarding. By the same token, John Unsworth has urged the creators of hypertext projects—electronic editions and archives—to document their failures "with obsessive care, detail, and rigor." Only with a clear sense of what has gone wrong and what can go wrong will it become certain that new knowledge is being evolved, whether about electronic editions and archives or about and entailing the electronic documents they contain.[20]

This chapter is organized around a series of errors or pitfalls that have emerged to date among the varied uses of the World Wide Web. Some are more easily recognized as human errors than others, but that shouldn't matter. Like the Internet of 1854, the uses (and potential misuses) of the Web point gamely to a host of assumptions shared by users, and negotiated in the unceasing growth and variety of the network as a context for meaning and a medium of communication. Because they are process oriented at multiple levels, the errors I consider help to supplement any consideration of existing Web sites—"cool" or explicitly "historical" sites, for example. Considering groups of Web sites stands to offer a suggestive sense of the Web as a synchronic form, existing more in moments of access. By hinting at process, however, the study of errors offers the Web as it exists more certainly across time and amid the temporality of labor: the work of accessing and searching, yes, but also the differently capitalized work of scanning, programming, cabling, linking, writing, designing, citing, and so on. The temporality of labor broadens Alan Liu's (2004a, 393) "class concept" of knowledge work to include the offshore data-entry technicians who so recently invented the Internet of 1854, the microphotography technicians who invented it before that, and the printers of 1854 who, just as unknowing, also helped to invent the same marvelous chimera.[21] I will argue that far from making history impossible, the interpretive space of the World Wide Web can prompt history in exciting new ways. Sound recording certainly did the same thing too, though that has long been rendered invisible by the unattended norms that produced the intuitive facticity of records and recording.

Error: File Not Found

The most persistent and consistently recognizable error confronted by users of the World Wide Web today is of course Error 404. When a user selects an outdated hyperlink or mistypes a URL, indicating the address of a Web server but not a viable page, the result is a transfer status code, "Error 404: Not Found." Different browsers handle this code differently, and different sites and servers offer their own versions of an error page. Like the subtle confusion of agency that emerges in the association of OCR scanning and human reading, Error 404 does not specify who committed or what caused the error to occur. It implies a dizzying potential for mistakes—more than four hundred different kinds—but stops short of laying any blame. It answers a particular request with a denial that affirms the constancy and ubiquity of the Web's administration, which is at once authoritative and impersonal—a system of protocols, really, that is seldom acknowledged but always present. Corresponding error messages, like the prefatory announcements on early phonograph records, hail users individually but are not by or from anyone in particular. Even when a Web master or systems operator replaces the generic error message with a server-specific version, she does so as a ventriloquist, speaking with the impersonal, authoritative voice of Web administration itself—a voice that reventriloquizes the impersonal authority that has so long hailed and conscripted its subjects to the mediated public: "post no bills"; "all circuits are busy"; "stay tuned for more." The Web administrator might offer to send users to another page on the same server, might promote the use of a search engine, or—in the dark humor surrounding the U.S. invasion of Iraq in 2004, for instance—might indicate that weapons of mass destruction have not been found.[22] Such humor works according to its incongruity or impertinence, but also according to the dispersed, impersonal mode of address that familiarly renders "have not been found" and "cannot be displayed" with such neutral authority.

The constancy and ubiquity of the administrative authority implied by the Error 404 status code stand in stark contrast to the primary cause of this errant result: Error 404 proliferates on the Web because the Web is constantly changing; pages are moved or deleted and links go out of date. When Brewster Kahle began the Internet Archive in 1996, he reported that Web pages remained online for an average of 75 days without being changed, moved, or deleted. Another account based on data from 2000 claims that "the average life span of a Web page is only 44 days," and since then Kahle has been widely quoted as saying that pages now live an average of 100 days.[23] Whatever the precise figure, and whatever its rate of change, change itself is a paradoxically consistent feature of the World Wide Web. And that causes particular difficulty for the varied users and uses of electronic documents.

A document stands as a potential ally in explanation when it is cited as meaningful and pertinent, and thus invoked within the public of its potential circulation. But the fluidity of additions, deletions, and modifications to the Web has helped to put such commonsense notions of "standing as"—of citing, and of public and publication—into confusion.

Some Web-based documents are far less perishable than Kahle's average Web page, but Error 404 still occurs with predictable frequency. For example, researchers have examined all of the articles "accepted in 2003 by the communication-technology division of the Association for Education in Journalism and Mass Communication," checking to see how many URLs in their citations remained viable. They estimate a "half-life" of roughly fifteen months. That is, researchers predict that half of all the links cited in any one article will be obsolete after that much time. The comparable half-life for citations in abstracts on Medline, the National Library of Medicine's database, has been estimated at seven years.[24] The differences between forty-four days, fifteen months, and seven years can be explained by disciplinary and institutional infrastructure: medicine and the National Library of Medicine both encourage continuity in a way that pedagogical trends in journalism and the Association for Education in Journalism and Mass Communication do not, as well as in ways that cannot directly attend the extraordinary diversity of "average Web pages" more broadly. Half-lives in any case measure the rate at which relevant publication contexts decay, suggesting an erosion—however uneven—of the public that defines and is mutually defined by Web publication.

Interestingly, if medical documents persist for relatively long periods of time on the Web, documents more explicitly about the history of the digital medium itself may fall a lot closer to Kahle's average. Different attempts are being made to preserve digital resources, but results are far from certain.[25] Paul E. Ceruzzi's (2003, 302) authoritative *History of Modern Computing* explains, "By January 1993 [Marc] Andreessen and [Eric] Bina had written an early version of a browser they would later call Mosaic, and they released a version of it over the Internet." But Ceruzzi's accompanying citation demurs, "Some of Andreesen's postings on the Internet have been preserved in an electronic archive. Since there is no way of knowing how long this material will be preserved, or whether it will remain accessible to scholars, I have not cited it here" (303, n. 50). Paradoxically, "it's still too early" (406), Cerruzi says, to tell the story of the Web, but the persistence of Error 404 indicates that it may also already be too late. The W3C's "Little History" of the Web must be little partly for this reason.

Error: Incorrect Formatting

If expired links whisper "too late," then the evolving standards for citing electronic documents murmur "too early." How can students or scholars begin to make documents stand as allies when the simple mechanics of standing remain in question? At one level, the characteristic mode of citation on the Web is a matter of making hyperlinks and pasting in URLs, creating "bookmarks" or adding "favorites." But when users like Ceruzzi wish to cite documents with more precision or on paper, they face another daunting array of potential errors, judging at least from the tangle of citation formats recommended by successive editions of the style manuals that have been published to guide them. Different guidelines for citation suggest different postures for "standing as," and can therefore reflect different assumptions about matters of evidence and issues of concern. (Too early and too late: uneasy in the face of just this contradiction, I have cited electronic sources in my endnotes below, along with archival resources, but the references that follow them contain only print publications.)

The first manual to attempt a systematic set of guidelines was Xia Li and Nancy B. Crane's *Electronic Style,* published in 1993. It was revised as *Electronic Styles,* plural, in 1996, when its authors added suggestions for citing from the Web, letter E in this excerpt from their table of contents:

A. CD-ROM and Commercial Online Databases
B. E-mail
 a. Archived Works
 b. Real-Time Works
C. FTP
D. Gopher
E. HTTP
F. Telnet
G. USENET
H. WAIS

This breakdown captures something of the confusion that users of the Internet faced in 1996, even apart from the problem of citing electronic sources. Li and Crane divide documents partly by mode of discovery (WAIS, for instance), partly by modes of delivery (e-mail, say), and partly by storage medium (CD-ROM, for example). None of this makes much sense today. Categorizing the Web according to its transfer protocol (HTTP) keeps it parallel to the Internet's file transfer protocol (FTP), but does little to acknowledge the distinctiveness of the Web as networked hypertext. Like "Gopherspace," "World-Wide Web"—as it was then frequently spelled, without the definite article "the"—offered early users much more than an experience of data transfer.

Li and Crane (1996, xv–xviii) give "embellished" versions of American Psychological Association (APA)–style citations (1993 and 1996) and MLA-style citations (1996) for electronic documents. The authors' embellishments are avowedly provisional; "Let's at least put something out there for the masses to use," they are reported to have ventured, "We [on the Net] are in dire straits." One librarian who reviewed *Electronic Style* protested its expedient, inductive approach: "Why have the authors given no reasoning for their recommendations? Why have they not recognized that an entirely new medium of communication requires discursive treatment, and not prescriptive examples by rote? . . . We need a rationale for a new system, not simply a cookbook."[26] Questions, proposals, and further discussion continued to populate newsgroups, listservs, and other online forums, overwhelming observers, while publishers, the APA, the MLA, and others have been busy producing and revising style sheets ever since.[27] In his hilarious review of the latest edition of the *Chicago Manual of Style* (2003), Louis Menand (2003, 125) complains about "the aggravating business of citing a Web page," noting about such matters, "The problem isn't that there are cases that fall outside the rules. The problem is that *there is a rule for every case.*" Bleary-eyed students with term papers to finish are stymied by too many choices, the same kind of problem that today "makes a hell of . . . shopping for orange juice: Original, Grovestand, Home Style, Low Acid, Orange Banana, Extra Calcium, PulpFree, Lotsa Pulp, and so on."

Guidelines for citation have always been arcane and multiform. They develop over time, as *Chicago*'s fifteen editions demonstrate. They vary by discipline, as the differing requirements of the APA and the MLA attest. And within each variation they vary by format—separate guidelines for citing journal articles, encyclopedia entries, oral histories, monographs, television programs, and so on, now confusingly joined by electronic formats like e-mail and Web pages. Complaints about the inductive or cookbook, case-oriented strategies of the style manuals are really about the profusion of forms and media, but they also more particularly concern the rhetoric of the respective disciplines that have failed to exert transparent, deductive reason over—have failed to discipline—the same profusion. How can Li and Crane's "masses" or Menand's bleary-eyed coed hope to prosper, when the very authors and editors of so many manuals cannot dare to hope for complete and final intellectual control over the ongoing flux? Contemplating *Chicago*'s 956 pages, Menand concludes, "The *Manual* is not too long. It is not long enough. It will never be long enough" (126). Each new edition will address itself to more and more publication formats, more and more varieties of orange juice, while a total theory of documents and a final system for documentation remain forever out of reach.

Despite the implication of so many manuals of style, format is not—again—what matters about documents. (Here's where orange juice and documents are not alike.) Remember that users of ProQuest and the microfilm *Times* can still cite an image of the paper as if it were newsprint: byline, headline, *"New York Times,"* date, section, and page number. Similarly, the users of the RFC database on the Internet today cite transcribed or imported nroffed text files as if they were actually the earliest RFCs (see chapter 3 above). And scholars use hefty bound versions of Victorian periodicals without any more than passing acknowledgment of the ephemeral single issues that the Victorians really read. "We may want to resist this tradition," writes Margaret Beetham (1990, 23), "but we cannot escape it" entirely. Footnotes habitually foreclose or disavow the long histories—and the labor—of preservation and migration that have made works like the *Times,* RFCs, and *Blackwood*'s available in different formats, because those long histories are not as important to authors or their presumed readers as the standing of such works as allies amid the public of their potential circulation. And if such footnotes foreclose media history, the popular discourse that lacks footnotes altogether of course forecloses even more: most spectacularly, there is the huge and habitual failure of so many Web pages to note or nod toward origin or provenance for the images and other elements they contain.

What is it about some documents, like the *Times* or RFCs, that makes material format so obviously incidental to their standing as allies in explanation? Different texts or versions of the *Times* and RFCs function so well as evidence because each work exists already as public record, standing within a robust discourse that is familiarly "punctual" in Michael Warner's (2002, 97) terms:

> Public discourse indexes itself temporally with respect to moments of publication and a common calendar of circulation. [And] one way the Internet and other new media may be profoundly changing the public sphere is through the change they imply in temporality. Highly mediated and highly capitalized forms of circulation are increasingly organized as continuous ("24/7 instant access") rather than punctual. At the time of this writing, Web discourse has very little of the citational field that would allow us to speak of it as a discourse unfolding through time. Once a Web site is up, it can be hard to tell how recently it was posted or revised or how long it will continue to be posted.

Format takes second place to context, to the social imaginary of potential circulation that must rely on commonsense norms of publication, including the periodicity of periodicals, journals, and broadcast programming; the partly calendrical logic of film, album, and video releases; and the experience of editions (and remainders) as such.[28] According to this analysis, Error 404 and the perishable quality of the Web may not be the most important question for historians and archivists to tackle. More urgent may be the evolution of a shared sense of Web publication as an event that can reliably be located and experienced in time, without error or exception.

Some Web publications do come dated in reliable ways, like electronic journals, so many date-stamped bulletin board postings, and elements of the rapidly growing blogosphere. But the punctual logic these forms adapt from print fails or falters elsewhere on the Web, where perhaps only Google's monthly updates offer users a tacit knowledge of Web publication as an event that exists definitively in time. Indeed, dates of update seem to be the most prevalent dates on the network, begging a chronology that is ever refreshed and refreshable, but rarely anchored against an explicit calendar of publication or circulation. Many have called for a way to certify whoness on the Internet—with electronic signatures or proprietary watermarks, for instance.[29] Yet the problem of whenness looms at least as large.

The Wayback Machine at the Internet Archive (<www.archive.org>) does allow users to see the pages at a particular URL specifically as they were on different dates in the past, and thus would seem to confound the continuous present of the Web. The Wayback Machine grounds its users' experience of the World Wide Web in time, as if they were flipping through back issues of a periodical, only their search results are neither strictly periodic nor completely bound by temporal logic. Users type in a URL and receive a consecutive list of random dates at which Alexa Internet's Web crawler has captured versions of that page. The resulting pages—sans JavaScript and other dynamic elements—are presented as time-specific artifacts, but the Internet Archive cannot itself generate the punctual logic that Web publication resists or denies. When users view pages from the past, captured to the archive's present servers, the relative extent and completeness of each past page is never obvious. Where will the edges and the empty "data islands" of each past document on the present Web be found? "Remember what Yahoo looked like in 1996?" rhapsodized one early account, appreciating the naive, uncluttered look of the early Web portals.[30] But search the Web using the Yahoo portal from 1996 and the results produced are today's Yahoo results. This archive is shaped like a Möbius strip. The Internet Archive servers are both of and on the World Wide Web that their collections seek to document, and like the "least recently modified" Web page, there is something oddly and unidentifiably present about the past to which the Wayback Machine promises to transport its users (though, like ProQuest, the Wayback Machine remains a wonderful resource).

The question of Web publication as an *event*—that "most condensed and semantically wealthy unit of time" (Doane 2002, 28)—must be posed in relation to other forms of publication, other media, and the medium-specific sense in which events are made public on the Web. This connection between publication-as-event and events-made-public is not transparent but is crucial to the experience of media in time and therefore in history. Like the serial media that it partially incorporates or "remediates," the Web represents time and simultaneously produces temporalities for its users; it records and performs.[31] Recording and performance diverge, in most cases, in the time lag between saving data and browsing through it. One happens, and then the other happens. Recording and performance are distinct when users play an MP3 a QuickTime file, for example, or display electronic text. But recording and performance also converge in important instances, as in the varied constructions of "real time" for webcam feeds as well as in experiences of interactivity more broadly. Neither divergence nor convergence is simple. On the one hand, the temporality of the Web resembles what Mary Ann Doane (2002, 28) calls "the indeterminacy, the instability and imprecision of cinematic time," and on the other hand, it evokes what Mark Williams (2003, 163) calls "the oddly protean temporality of television," which is so marked by live broadcasting.[32] The latter case, at least, may appear to have little to do with electronic documents, but real time on the web both represents and produces the instantaneity or the sense of present into which documents are published, and thus it helps to structure publication as an event.

Like television's *liveness,* however, the *real time* of the Web is an intricate construction. "Both terms are grounded in the capacity for electronic media to represent something at roughly the same moment it occurs," explains Williams (2003, 163), "But each term, in significantly different registers, also designates a key dynamic of disavowal, in that each names an act of mediation but also the desire to experience this act as unmediated" and unlabored, immediately lived and immediately real. Real time is more of an effect, then, an experience of data "on the fly," than it is the literal copresence or cotemporality of users and events. And the instantaneity that real time represents and produces is not always exactly instantaneous either—at least not yet. The real-time media effect is "propped onto the near future," Williams says: it offers a sense of present that depends in part on the fattest possible bandwidth and fastest possible downloads, which remain more dreamed than everywhere accomplished (163). Soon but not yet is the eternal promise, evident in the jerky miniatures of QuickTime "movies" that "struggle against (as they struggle to become) cinema," according to Vivian Sobchack (2004, 307); and evident even more pervasively in the intermittent delays of Web surfing, where traffic on the network and files of various sizes result in unpredictable load times and varying degrees of user aggravation—at least for now.[33] "Online temporality," Alan Liu (2004a, 225) notes, "amounts to antidesign" because of this unpredictability.[34]

If the real time of the network were real, every click would instantly gratify, since clicking is the user's most certain experience of an instant in time. Each click is performed by user and machine together at their point of closest physical contact—digit to digital—and those present instants are simultaneously recorded into "history," according to the parlance of the Mosaic browser and now of Microsoft.[35] Like the "back" command all browsers share or the "undo" command editing programs often include, Internet Explorer's "history" button chronicles consecutive instants in time, distinguishing them as places, sites of performance, once present and now past.

Documents can be published into the instantaneity of real time on the Web in fairly short order, with a few simple clicks: it will take "about 10 minutes worth of work, not even," explained the project manager of the William Blake Archive to the project's editors in July 1997, for example. The William Blake Archive is a project that began in June 1995 to publish digital facsimiles of Blake's work along with tools and apparatus to support Blake scholarship. Editors and programmers working with or at the University of Virginia's Institute for Advanced Technology in the Humanities (IATH) labored steadily, designing the archive, writing code, collecting and marking up text, and scanning and tweaking images. After two years, they were finally ready to publish their first Blake work: an electronic edition of a single copy of Blake's *The Book of Thel,* known as Thel F. "Publication entails two sets of actions," wrote the project manager in an e-mail to the project's private list; "I need to create live links to the Search page, Thel F itself, the bibliographies, and the new archive update, and drop the warning screens down in front of" the pages still not ready for publication, and "we need to publicize the Archive by sending copies of the update to relevant listservs."[36] The first task was accomplished Friday afternoon, August 1, and the second waited until Monday, August 4. The electronic edition of Thel F was published and publicized.

Error: *Private* and *Public*

The William Blake Archive is likely one of the most exacting and well-documented publication projects on the Web as well as one of the most self-conscious, so it offers a rare opportunity to look inside the publication process and think about publication as an event. Whether it is properly an "archive" may be debated, but the William Blake Archive resonates with questions of preservation, preservability, and dissemination that the previous chapters have encountered in relation to both voices and thoughts from the past: voices "captured" onto tinfoil and wax; thoughts expressed typographically in critical editions or alphanumerically in speculative libraries of the future. As a noncommercial publication project connected with a state research university, the William Blake Archive helps to promote the Web as a public arena for what IATH and others have termed the "digital humanities." This distinguishes it from so many commercial uses of the Web today, but both commercial and noncommercial users with a Web presence must depend on acts and events of making public.

Clearly, the publication of the William Blake Archive "Thel F: The Electronic Edition" was massively "front-loaded"; it depended on years of labor by a team of scholars—the results of which only fully became public with the creation of "live links" between the existing William Blake Archive Web page, Thel F, and its accompanying apparatus. All publication forms are front-loaded—think of writing and editing a novel before it gets sent to press, or shooting and editing a movie before it goes to theaters—but the transition from private to public is rarely this easy. Publishing on the Web is so easy, in fact, that it sometimes happens accidentally—for example, when a designer for CNN's Web site unintentionally published links to the in-progress obituaries of Vice President Dick Cheney, Fidel Castro, Nelson Mandela, and other still-living public figures. The obituaries were on an internal, private site that was not password protected, so publishing the URL was publishing the page. In this case, the event of publication and the publication of events together proved untimely.[37]

Like those pending death notices, the William Blake Archive Thel F was already there, already present on the IATH servers, and it was already named on the William Blake Archive Web page, only now it was called into public through the act of linking, where it could be publicized to the "relevant listservs" that were presumed to attend and partly to produce the public of its potential circulation. Thel F would then produce its own public too, which the main archive page addresses with an emphatic "WELCOME," whether users visit "for pleasure, study, or intensive research." That public has emerged gradually since 1997 according to interest and opportunity, but also according to rules for reasoned behavior in public. The main archive page outlines a set of conditions—albeit largely unenforceable—for use and reuse of its contents, and it warns, "By accessing the Archive, you acknowledge that you have read and accepted these conditions."[38] Merely to access Thel F is to subscribe to the decorum its editors specify.

As an explicit extension of that decorum, several months after the publication of Thel F, the editors amended the main archive page to include a preferred citation format, formalizing the identity of the edition comprised of documents as follows:

> Blake, William. The Book of Thel, copy F, pl. 2. The William Blake Archive. Ed. Morris Eaves, Robert N. Essick, and Joseph Viscomi. **13 November 1997** <http://www.blakearchive.org/>.

Though it probably represents the date this example was composed, the November date (emphasis added) included in the sample citation is supposed to be a date of access, according to the editors' instructions, to be rolled forward by any user of Thel F, plate 2. (Like Licklider's hypothetical session on a procognitive system, the William Blake Archive editors interpolate their own present into a speculative future.) Neither the date of Blake's original printing in 1795 nor that of the "live links" established on August 1, 1997, forms part of this citation. The URL points to the main archive page, associating the digital image of plate 2 with the larger context of its electronic publication. That context appears both in-progress and eternal, publishing Thel F and other documents into an instantaneity that extends "for the foreseeable future" (Kirschenbaum 1998, 239), and welcoming users to the archive in and into a continual, continuous present tense.[39]

The continuous present tense within which the William Blake Archive and similar editions exist on the Web resonates less with the instantaneity of the real-time effects that attend them than with the cultural logic of timelessness, whereby canonical figures like Blake stand as rich and enriching subjects of inquiry. Like Edison's Washingtons, Lincolns, and Gladstones (1878), that is, or Project Gutenberg's Declaration of Independence (1971), and the continual flogging of "opera" records by the early-century phonograph companies, the William Blake Archive is helping to make a new medium authoritative in a sense by co-opting cultural authority, by entwining the new means and existing subjects of public memory (as well as, of course, by invoking the authority of institutions like IATH, its sponsors, and the participating Blake repositories). But the present of Web publication in general can hardly resonate with the alleged timelessness of "the classics" when so many of its subjects are noncanonical, commercial, or banal. Its continual continuousness must be plumbed beyond the subjects of its pages.

Lacking the cultural, institutional, and financial capital of outlets like IATH (or CNN), everyday users of the Web today create Web pages by themselves, using one or another of the available off-the-shelf, WYSIWYG editing programs, like Microsoft's FrontPage or

Macromedia's Dreamweaver, and then posting the results to a commercial Web server. In his analysis of the "interface metaphors" that characterize Macromedia products, Tarleton Gillespie (2003, 119, 115, 113) notes some of the ways that users, aware or not, are constrained by the terminology, defaults, and menu options built into such software. One telling—if idiosyncratic—result of the constraint they experience is the frequency with which users have accidentally published Macromedia examples. Dreamweaver includes a tutorial that walks users through the creation of a Web site for an imaginary coffee shop named Scaal. Gillespie used a search engine to look for *scaal* and found hundreds of sites called "Scaal Home Page." "A few were exact copies of the Dreamweaver tutorial site," he explains; "presumably, someone was practicing and inadvertently posted the pages to their public server. The majority were actual websites, for different products and interests; it was clear that these users had generated their sites by modifying the HTML (hypertext mark-up language) code from the tutorial, but had failed to change the title of the homepage." These imaginary coffee shops lie scattered like ghost towns in the continuous present of the Web, sometimes a whole shop, but usually just a leftover sign hanging on top of a window.

Like CNN's premature obituaries, the Scaal sites again affirm the relative ease of publication on the Web, where private and public coexist, and the distinction between them is not intuitive, not present, for all users at all times: the present presentness of Web publication varies according to users' attention and expertise. More important, though, the Scaal sites suggest that data and metadata exist at different registers of public and publication. That is, users publish their own pages called Scaal Home Page when they alter the <body> the tutorial gives them, but leave something of its <head> unchanged. Encouraged—that is, constrained—by user-friendly WYSIWYG editing tools, users have accidentally distinguished the publication of their page from the publication of its encoded self-description. Like so many other errors, their mistake points toward underlying assumptions and unanswered questions that attend electronic documents more broadly.

All digital objects contain data and metadata. Corporations and allied institutions usually have much more invested in metadata than individual users do (users of Dreamweaver can largely ignore markup, for example, because the program modifies HTML tags to reflect their design choices), but metadata is always present, whether the digital object in question is a packet, file, message, or page. Even tangible digital objects, storage media like diskettes and DVDs, require metadata. All of the information on a DVD that users watch are data; the information they do not watch are metadata. Some metadata become visible in menus and titles, but a lot more remain unseen. For instance, every DVD includes an arbitrary chunk of information about its anticipated location, since the Motion Picture Association of America made manufacturers divide the globe into six regions. As part of corporate efforts to control piracy, DVD players are only able to play DVDs from their own region. Are those chunks of information literally part of the "movies" that users watch? The question may seem pointless to movie fans, but related questions about the relationships of data to metadata have consumed scholars who design and publish electronic archives and editions. What is "the conceptual status of markup?" Dino Buzzetti probes; "Is it a sort of metalinguistic description, or is it a direct expansion of our writing system employed to express intrinsic features of textual content?" At stake is the very being of "the 'text' itself." Either markup is not part of the text or it is.[40]

Scaal errors notwithstanding, it would seem that publishing a Web page or related digital object is publishing its markup, rarely viewed but always there. Data and metadata are inseparable. The facsimile images published within the William Blake Archive, for instance, each come salted with textual metadata, bibliographic information about the original engraving that the image represents, and information about its production process: the date the image was scanned, from which source medium, and with what technical specifications, hardware, software, file size, resolution, and so on. This information travels everywhere the image does. Though "an image file is typically thought of as consisting of nothing but information about the image itself—the composition of its pixilated bitmap, essentially" (Kirschenbaum 1998, 240)—these JPEG files consist in part of carefully detailed production records, which remain hidden from view unless interested or particularly "responsible" users go looking for them.[41] But if these textual metadata always adhere backstage, in effect, they can also always be redescribed by the editors and programmers to contain more or different information. Two months *after* they published Thel F in August 1997, for example, the editors worked on revising the markup terms and templates that pertained.[42] Today, the electronic edition of Thel F is nowhere dated 1997; it is dated summer 2000 on a "Revision History" page that identifies the season of its "conversion to Blake Archive Description DTD 2.1." For now, at least, the August 1, 1997, publication of Thel F may only be documented as a discrete event in the history of the World Wide Web according to archival copies of the editors' private e-mails, which can be had either by soliciting and arranging access to private electronic files, or by flying to Minneapolis to consult copies that were printed out and deposited in a conventional archive, processed and preserved in boxes by the Charles Babbage Institute at the University of Minnesota.[43]

The "Blake Archive Description DTD" defines the markup elements for documents specific to the William Blake Archive. "Thel F: The Electronic Edition"—comprised of data and metadata—was published and then revised, when editors and programmers added new fields for more metadata, and made adjustments to the relationships between data and metadata. Revisions to the DTD were incremental, proceeding as part of the work of editing different Blake texts, but once a substantial number of changes had been made, the published Thel F data were converted to the new version, DTD 2.1. Here, for instance, are the opening lines of the DTD published by Jerome McGann (1996, 159–160) for documents in the Rossetti Archive, another project at IATH:

```
<!—This is the DTD for the Rossetti Archive document
(rad) structure—>
<!—revised: 6 Oct 94 to add titlePage tags (seg)—>
<!—revised: 9 Mar 95 to add r attr to l, lg and lv
(seg)—>
<!—revised: 25 Apr 95 to add gap and orn.lb tags
(seg)—>
. . .
<!—revised: 15 Jan 96 to change commentaries to
generic sections
and "." style names to Caps style
and rad to header and text (including group)—>
```

As McGann (2001, 91–94, 13) explains, not every change to the Rossetti Archive is reflected here, and some of these revisions are more significant than others. The important revision of March 1995, for instance, introduced a scheme for collating Dante Gabriel Rossetti's texts and automatically identifying variant versions among multiple documents. Each revision, however incremental, better represents or allows access to Rossetti's own "obsessive" revisions.

My purpose is not to belabor the arcane, or intervene into ongoing discussions of SGML, XML, or specific formats and encoding strategies, but rather to elaborate some of the complexities of Web publication as an event. Not only do Web publications in general appear to compromise or eschew the punctual logic of more conventional public discourse but the different components those publications possess are differently public: data and metadata are at once mutually copresent and versioned according to separate calendars. And since metadata are precisely "meta-," their revision process reflects an author's, editor's, or programmer's ongoing reinterpretation of the data in question. Users might alter or correct a <head>, for instance, to better fit their evolved understanding of its <body> or the relationships that headers and bodies share. This is what makes an electronic archive or edition, and ultimately—I would argue—the World Wide Web itself, into what McGann (2001, 11) calls "a machine for exploring the nature of textuality." And once the machine is running, it becomes all but impossible to experience Web publication as a bounded event within a punctual discourse. Even the most exactingly self-conscious, intensively authored and attended publications on the Web, those scholarly electronic editions and archives, tend to exist in a sea of installments, versions, and revisions, works continuously present and yet constantly subject to change. Versions and revisions—long the raison d'être of critical editing—are now also among its ongoing and unapologetic results.

However marked the World Wide Web may be by dates and updates, by versioning and revision, however pocked it may be by expired links, and however haunted by the promise of "soon, not yet," posting something on the Web today means publishing into a continual, continuous present that relies more on dates of access and experiences of "WELCOME" than on any date of publication. Increasing commercial uses of the Web only intensify that presentness: "Buy now!" so many pages welcome, updated with avidity for one-click shopping and real-time trading. But the World Wide Web is more consistently a text than a market ("The Web is a written place," as David Weinberger [2002, 165] puts it), and the continual, continuous present tense of Web publication in this respect must be described with care. It has a rhetorical force and also harbors significant economic implications. The present tense of the Web defies the traditional logic of intellectual property law, for one, which is based on monopolistic rights offered to creators for limited terms. As Lawrence Lessig and his colleagues protested—unsuccessfully—in *Eldred v. Ashcroft* (2003), Congress has muddied the logic of limited terms for copyright by retroactively extending them.[44] The present tense of the World Wide Web in this sense suggestively coincides with and may even buttress a new logic for authored commodities—a logic less atemporal than it is antitemporal, subject to the redefinition of limited terms and limitability by legislators responding to pressure from a corporate class of owner-elites who increasingly see "*content* as a strategic corporate asset."[45]

But the present tense of the World Wide Web follows an ancient rhetorical tradition at the same time that it does the febrile logic of "late" capitalism and global finance. This is the present tense of hermeneutics, writing about writing, and interpretive processes more broadly. It is the present tense of dream interpretation (Joseph to the pharaoh: "The seven fat cows are seven good years . . ."); the present tense of citation, quotation, and cross-reference ("The author notes on page ten . . ." and "See also . . ."). It is the present tense of slide shows, photo albums, and scrapbook exhibitions ("Here I am at Lake Michigan"). It is the present tense where fictional characters live: Odysseus sails; Hamlet soliloquizes; Dorothy clicks her heels. It is not—to distinguish—the present tense where plants and animals live ("Chrysanthemums require long nights to flower" and "Beavers build dams"), not the present of nature and modern science ($e = mc^2$). Nor is this the insidious ethnographic present tense of early anthropology, whereby Western observers chronically "deny the coevalness" of others, as Johannes Fabian (1983) explains. Far from denying the coeval, the World Wide Web produces coevalness according to the singularity, plenitude, and instantaneity of its interpretive space. In short, the Web offers a space for interpretation where interpretation is always already underway; the machine—a disciplinary machine proper to the humanities—is running, whether users acknowledge it or not. "This is a work in which anyone can join," might be its slogan, where the figure of "anyone" produces and is simultaneously produced by the emerging new media public in which interpretive work, like battle, is continuously joined by users who browse and click as well as by those who cut and paste, post and publish.

The phrase "work in which anyone can join" comes from the original call for contributors to the *Oxford English Dictionary* (OED) in 1879, and the dictionary offers a helpful point of contrast and comparison. Unlike the World Wide Web, of course, the OED was produced with a central, institutional authority—James Murray and the Philological Society of London—and organized according to a preexisting scheme—the alphabet. Like hypertext, though, the OED publishes a web of excerpts and references, dated quotations to illustrate evolving usage. That singular, immense web, Seth Lerer (2002, 109, 108) explains, "was built collaboratively, out of Victorian habits of reading," as scattered members of the English reading public sent the editors thousands of slips of paper bearing the examples of usage they encountered. And as such, the dictionary assembles "not only a history of language but a history of reading taste, a record of the ways in which the scope of English prose and poetry was understood" in the decades of the OED's compilation. One might say by comparison that the World Wide Web publishes a web of excerpts and references, pages and links, built collaboratively by the euphemistic "anyone" of contemporary computer literacy, Internet access, software and hardware resources. But imagine an OED in which none of the quotations are dated or, better, no single understanding of their datedness holds sway. Such a dictionary could offer no history of language, but it could still encapsulate and thereby document the reading practices of its own "anyone" for the period of its compilation. If that compilation were ongoing, like the

World Wide Web or the new edition of the *Online OED,* then it would continuously encapsulate the reading practices and interpretive strategies of a continuing present.

This is something of what I mean when I call the Web an interpretive space in which interpretation is always already underway, or a machine proper to the humanities. Both metaphors, like the OED analogy, imagine a public variously engaged in reading, selecting, excerpting, linking, citing, pasting, writing, designing, revising, updating, and deleting, all within a context where the datedness of these heterogeneous *interpretive* acts remains inconsistently perceived or certain. Just as an OED without datedness would no longer offer a history of language, the World Wide Web offers scant help on a history of "the language of new media," in Manovich's phrase. Users catch glimpses—via the Wayback Machine, for instance, and in the litany of revisions noted in a DTD, in old threads of discussion available via Google Groups, in the archived entries of the blogosphere, or in the obsolete RFCs still offered for "historical purposes"—but users must interpret each with care. Nostalgia—a close relation of Vivian Sobchack's (2004) "mnemonic aesthetic"—encourages varied "emulations" of the past, like the browser emulations at Deja Vu (<www.dejavu.org>), which allow users to experience today's Web as it might have looked using Netscape Navigator 1.0 (1995) or Internet Explorer 2.0 (1995). Indeed, even apart from nostalgia, emulation is one possible preservation strategy being discussed by specialists interested in archiving digital resources.[46]

It is not that the Web resists history per se, only that electronic documents compel attention to themselves as differently—often dubiously—historical, where history always happens at the levels of—at least—data, metadata, program, and platform. (Think of the open source movement and the incremental versioning of Linux.) Seen in this light, the Web presents an opportunity at least as much as it presents a problem. Amid an unrelenting contextual historicism—the developmental arc or time code of "late" capitalism along with the unshakable, concomitant ideology of progress—the Web helps to pose the question of history itself. Electronic documents may defy bibliography, but they inspire keenly bibliographic impulses. Each lacuna in provenance (the discomfort of not knowing where a digital object comes from) can help put provenance on the table. I am being sanguine. The putative "end of temporality" associated with today's communication technologies need not mean the end of temporality any more than the nineteenth century's telegraphic "annihilation of space" put an end to space.[47] Emulation works as a preservation strategy in part because it helps self-consciously to underscore the differences between pages and documents—that is, between issues of format and matters of concern. In this respect, the World Wide Web offers a vernacular object lesson in what Jerome McGann has called "the textual condition." As much as the authenticity of original documents serves as an intuitive source of orientation and in some contexts (like fiduciary institutions, state bureaucracies, and the courts) a matter of practical necessity, the original as such does not exist, since documents are "only" social experiences of meaning. The document I write and the desktop window I write it in are not materially different from one another. Both are data and metadata saved to the hard drive, and represented on the screen. The former stands as a document because of its context, not its body, where context involves the whole social realm and human labors of literacy, inscription, writtenness, and computing; of representing and saving; that is, of meaning and the presence of meaning.

The H-Bot

None of this is to deny that the World Wide Web presents interesting challenges for media history, or that U.S. culture today continues to enjoy a vexed and varied relationship with historical memory. A historian at George Mason University is working with the Center for History and New Media there to develop a program he calls "H-Bot," a personified search engine designed to liberate students from the tedium of memorizing dates. It is not a mnemonic device to help users remember but rather a mnemonic prosthesis to harvest dates from the Web so users don't have to remember. Like the introduction of pocket calculators in math and science classrooms, the H-Bot is intended to spare students from relative trivialities so they can attend to more interesting and conceptual stuff.[48] At present (in "alpha release"), the search engine greets users on the Center for History and New Media site:

> I'm ready to help you find a year in which something happened (between 1000 and 1990 AD), and possibly the exact date as well (if applicable). **Please type in your search phrase in the past tense and use details where possible.** For example:
> **» Charles Darwin was born (rather than "Darwin was born")**
> **» The Magna Carta was signed (rather than "The Magna Carta")**
> **» The Berlin Wall fell**
> **» Queen Victoria ascended to the throne**
> **» The Battle of Hastings was fought**
> **» James Joyce's Ulysses was published**

Users enter a statement into a query box and click the button marked "in this year, . . ." Relying in part on the analysis of results produced by Google's search algorithms, and in part on local databases and standard query language, H-Bot identifies the dates events occurred. For instance, in 1.9 seconds it can reply to the statement "James Joyce's Ulysses was published" with the rejoinder "I am extremely confident that the year was 1922." Repeated searches take different amounts of time, depending on server and network traffic as well as the specificity and decorum of the query. The H-Bot is unable to answer questions about events that happened in more than one year like "James Joyce lived in Dublin" (a range of years) and "Eugene V. Debs ran for president" (several different years). Specificity in phrasing queries makes a big difference. And H-Bot is capable of some interesting errors.

H-Bot can sometimes mistake fiction for fact. Asked when martians landed in New Jersey, for instance, the H-Bot is extremely confident that the year was 1938, the year of Orson Welles's "War of the Worlds" broadcast. Asked when Leopold Bloom walked around Dublin, the H-Bot is extremely confident that the year was 1904, the date of events represented in Joyce's novel of 1922. Both examples arise from the relative lack of distinction among representations on the Web: events are events, whether they actually happened or not. The H-Bot does not distinguish martians or Bloom as fictions because it does not recognize the multiple levels at which interpretation operates—a matter of semantics that reaches far beyond local or grammatical distinctions between the present tense and the past. Asked when the telephone was invented, the H-Bot is extremely confident of 1860, because of the work of Antonio Meucci and possibly Philip Reis; asked when Alexander Graham Bell invented the telephone, and the answer is 1876. The normal meaning of *invented* makes these contradictory results, but the H-Bot is blind to this semantic issue as well. Its extreme confidence comes not from any real understanding or artificial intelligence but from the native assumption of its users that a perfect search algorithm can someday be developed to analyze Google's index of the World Wide Web, which is structured in part according to Google's own problematic assumptions about what its designers call the "uniquely democratic" nature of the World Wide Web.[49]

Like the Internet of 1854, the H-Bot gets things "wrong" only to the extent that users allow its personification to displace or repress their attention to the aggregate human labor according to which the program actually functions. The H-Bot works exactly as programmed on data precisely as they stand. Its errors are slips of the tongue, occasions for users to glimpse again their own unflagging and largely unconscious desire for reading machines and self-knowing texts, for what Tim Berners-Lee and the W3C actually already envision as the Semantic Web, a better, increasingly self-reading version of today's World Wide Web.[50] The H-Bot in its later incarnations may well circumvent the martians in New Jersey and other such chimera, but I hope it doesn't. As a pocket calculator for dates, the H-Bot stands partly to reinscribe the limited construction of historical events that its designers seek to trivialize: births, deaths, battles, treaties, and ascensions to the throne. But as a game for trying to generate chimerical results, the H-Bot challenges players to think about historical events as well as digital representations as fully complicated interpretive acts. In what sense do fictional events "happen" on publication? In what ways can the definition of *invented*—or of *the* telephone—be interrogated and enlarged? How might events themselves be produced by the retrospective inquiry that interprets them *as* events? Players test the H-Bot's "understanding" by self-consciously honing their own knowledge of history, the Web, and history on the Web. In short, history is the object of this game. As the object, history requires, first, a significantly detailed knowledge of the past; second, a modest sense of how search engines function; third, a broadly speculative sense of the kinds and variety of representations of history on the Web; and fourth, a related sensitivity to "history" as itself historically produced. Players posit the accumulated labor that has produced the searchable Web, and they enjoy—they win—a richly historicized version of history.

Epilogue: Doing Media History

Good words do not last long unless they amount to something.
—Niimiipu Chief Joseph, Washington, DC, 1879

In 1877, while Edison was busy working on telephones, telegraphs, and tinfoil phonographs at his laboratory in Menlo Park, New Jersey, the United States was still fighting Indian wars. In October, the military campaign and subsequent retreat of the Nez Percé came to a disastrous end in what is today the state of Idaho. Chief Joseph surrendered to U.S. Army general Howard with the immortal words, "From where the sun now stands, I will fight no more forever." His words were immortal according to the offices of Arthur Chapman, an interpreter, and Charles Erskine Scott Wood of the general's staff, who carried a pencil and a pad of paper and was able to collect them as they "fell from the lips of the speaker." *Harper's Weekly*—the self-avowed "Journal of Civilization"—published its version of Wood's account in November 1877 for general circulation. Wood later gave away his manuscript, which then disappeared, and made a copy, also lost, but he was still memorializing the fugitive's surrender inside quotation marks as late as 1936. Chief Joseph, meanwhile, told his own version of events to a reporter in Washington, DC, where he made his case to federal officials. The *North American Review* published Joseph's account under the title "An Indian's View of Indian Affairs" in April 1879. It details the iniquities suffered by the Nez Percé at the hands of government agents, up to and including the lies that prompted Joseph to surrender that October, with the immortal words, "From where the sun now stands, I will fight no more."[1]

Which did Chief Joseph say: "I will fight no more," or "I will fight no more forever"? Strictly speaking, he said neither and he said both. He did not speak words in English, and the "no more" version nests within "no more forever." As a matter of public record, however, Joseph's words remain indistinct for other reasons as well, because documentary evidence is scant, and because no sound recording exists or could exist. Check Google today, and the jury is still out. "I will fight no more forever" is the most common version, but "I will fight no more" and "I will fight no more against the white man" both get "votes," according to Google's organizing metaphor of linking as voting. Knowledge of sound recording and the familiar certainties of schoolbook history—"just the facts"; "true or false"—help, on reflection, to make this a pretty uncomfortable situation. Neither, both, and one-or-the-other: unless we do successfully "provincialize Europe" and critique empiricism, in Dipesh Chakrabarty's terms, Niimiipu chief Joseph must have had some exact words. The practices of sound recording play a forgotten role in this discomfort, because at a basic level speech is made public and falsifiable or "exact" according to the offices of its imagined and culturally imaginable mediation.[2] Speech gains immortality, that is, partly according to all of the constructed instruments and institutions of its potential preservation. The *forever* that attended and/or appended Joseph's surrender serves as much to address the problematic involved in collecting and preserving speech as it "falls from the lips" as it does to suggest the finality of his capitulation.

The continued indeterminacy of Joseph's speech hints anecdotally at the particular conjunction of interpretation and preservation—of records and documents—that I have been addressing at length in these pages. Joseph's or Wood's *forever* rings with the question of its own immortality—a question that cannot be divorced from the interconnected subjects and instruments of inscription as they have variously been deployed and experienced. Google hits on Chief Joseph's "fight no more" suggest the latest and now digital iteration of the long-lived romantic tradition of Indian oratory in the United States, wherein the characters of John Logan and Red Jacket "spoke" stirring words from the pages of schoolbooks, available for memorization and recitation—for hits, one might say—by generations of U.S. schoolchildren. Only with the burden of contexts and questions such as these can Edison's claim to have "captured" "fugitive" sounds for the first time in 1877 read clear.[3] And only with the attendant depth and complexity of *forever*—in or of Chief Joseph's words, neither and both—do the keywords *record* and *document* really make sense.

In the preceding pages a different yet related *forever* flits behind the scenes, emerging here and there in fleeting allusions to literary history, of which the romantic construction of Indian oratory in one sense forms a chapter. Because of my own training and experience, literary history has been my ready example of the humanities, that group of related disciplines that emerged institutionally toward the end of the nineteenth century with the "peculiar burden," Lawrence Veysey (1979, 52) writes, "to represent the heritage of higher 'civilization.'" My object has not been to solidify some point about the literary as such, or about "civilization," certainly, but rather to promote by example the ways in which doing media history can grant partial access to the epistemologies and practices of humanists and the humanities. As Lorraine Daston (2004, 363) has observed, there is a healthy and diverse literature on the sociology (for lack of a more comprehensive term) of scientific knowledge. Scholars have considered "how biologists learned to see under the microscope, how botanists learned to characterize plants in succinct Latin, [and] how physicists learned to abstract from messy phenomena to mathematical models." Historians of science have offered their readers "a social history of truth" itself.[4] But far more rare are considerations of how knowledge in the humanities comes about: How have literary critics learned to criticize? Asks Daston (363), "How do art historians learn to see, historians learn to read, philosophers to argue? What is the history of the art-historical slide collection, the initiation into archival research, the graduate seminar?" What are the sociological origins of truth in the humanities? Media history bears closely (though not exclusively) on questions such as these. Better still, media history offers access to the epistemologies and interpretive practices of the humanities at a vernacular as well as scholarly or academic level. Media aren't the instruments of scholarship in the humanities; they are the instruments of humanism at large, dynamically engaged within and as part of the socially realized protocols that define sites of communication and sources of meaning. Media history offers nothing less—if also a great deal more—than the material cultures of knowledge and information.

How does the early history of recorded sound connect to Moses Coit Tyler's collation of early American literature? How does the early history of distributed digital networking connect to National Endowment for the Humanities–funded editions of Emerson and other literary lions? I have tried to make gentle, contextual connections, suggesting that media history and literary history share the same groundwater, not that one drives or determines the other. By implication, one of the great benefits of doing media history is that it latently offers what anthropologist Alfred Gell (1992, 42) calls a "methodological philistinism" with which to glimpse the broadly canonical cultural productions—literary and otherwise—and canonizing operations that inform humanism and humanistic inquiry today. Gell uses an analogy to describe his perspective, noting the distinction that routinely and intuitively gets made between studying religion while putting aside questions of belief (anthropology of religion, religious studies) and studying religion while believing it (theology). One might say by comparison that there must be a similarly valuable distinction to be drawn between studying literature while putting aside questions of aesthetics (media history) and studying literature while "believing it," naturalizing literariness and literary aesthetics (English).[5]

Let me quickly emphasize, as Gell does, that doing media history does not make one a philistine; rather, it can offer a methodological detour around the aesthetic in order to make the multiple conditions of its cultic status (that is, aesthetic value) more clear. The goal is to understand the aesthetic in a broader, more catholic sense.[6] This amplifies a move made so deftly by John Guillory (1993, xiii) in *Cultural Capital,* which aims at today's crisis in the humanities by focusing "debate away from the question of who is in or out of the canon [and toward] the question of canonical form in its social and institutional contexts. The form we call 'literature' organizes the syllabus and determines criteria of selection much more directly than the particular social biases of judgment which have been invoked to explain the canonical or noncanonical." The social and institutional contexts that have produced literariness as a canonical form—as a belief, Gell might say—include as well as partly consist of media, the techniques and conditions that structure communication as cultural practice, and that thus provide the ground for any specifically literary communication, whether the early American texts established by Tyler or the U.S. authors edited by the Center for Edition's of American Authors with the imprimatur of the MLA. As Guillory puts it, drawing on Walter Benjamin, "Canonicity is not a property of the work itself but of its transmission in relation to other works in a collocation of works" (55).

Guillory's idea of transmission interrogates "the syllabus in its institutional locus, the school"; although one might well probe more generally, to interrogate "the ways in which knowledge has been, is, and will be shaped by the transmissive means through which it is developed, organized, and passed on." The "arts of transmission" so designated embrace the history of the facticity of the modern text speculated here (see chapter 3)—including the histories of—writing, print, and nonprint media as they have developed and continue to develop in mutually defining ways, as foils.[7] It is in this spirit, for instance, that Meredith L. McGill's (2003, 8) history of reprinting in the United States (1834–1853) shows how disputes over intellectual property rights helped "to structure the literary field, and how the question of the cultural status of the literary [got] folded into the texts themselves." It is in a similar spirit, further, that Jorge Cañizares-Esguerra (2001, 6–8, chapter 1) connects baroque debates about Amerindian scripts and the history of writing to the development of a modern, historiographical sensibility more commonly dated to the end of the eighteenth century. Like McGill's literary field or Cañizares-Esguerra's historiographical sensibility, Tyler's records and Emerson's documents have been structured in part according to the wider economy of records and documents—aesthetic or nonaesthetic, "literary" and not, inscribed on paper or elsewhere—within which each came and continues to enjoy cultural currency within the contexts of its own *forever.* Disputes, debates, and economies broadly proper to the history of media have helped variously to produce the literary, the historical, and the like—to produce, that is, the data of culture as such.

What are the contexts of *forever* today, now that so many channels of communication are digital, and now that the data of culture are increasingly articulated, processed, transmitted, and stored electronically? The foregoing pages suggest how broad the relevant contexts are, while the ongoing crisis in the humanities indicates just how unsettled and at times contested they can be. Whereas the new medium of recorded sound emerged out of and into the chaos of industrialized communications at the end of the nineteenth century, new media today emerge out of and into a comparable chaos called "the postindustrial." Recalling the nineteenth-century version, this chaos entails that provenance is routinely in question (bibliographically), that reception is frequently in doubt (sociologically), and that authorial ownership is keenly in dispute, as the global marketplace rushes on and roils the law. What are the political economies of writing and reading, of seeing and knowing, online? How should electronic information be secured and preserved? Who knows where this or that digital content comes from? Which images have been doctored? Who sent me this spam? New media emerge amid the chaos that they help partly to reconstruct as order, the so-called logic of the postindustrial and postmodern.

It is tempting to see the two cases elaborated here as indexes of vaguely symmetrical epochs in the United States: on the one hand, the "Search for Order" (historian Robert H. Wiebe's description of 1877–1920) pursued at the expense and to the exclusion of Chief Joseph, among many others; and on the other hand, the "New World Order" of single-superpower status pursued at the expense of denominated "evil ones" and attendant "collateral" casualties, for instance, in places like Iraq. But questions of Order (with a capital O) at levels like these are well beyond the immediate scope of these pages. Media history offers instead the always emerging "order" (with a lowercase o) of public life and public memory. Even if it can be difficult to read lowercase order into the big picture, quandaries and contests over the meaningfulness of specific inscriptions suggest that such connections do exist, and that they must be plumbed with care. My examples may be phonograph records and electronic documents, but one might think as well of draft cards, green cards, and other paperwork, for instance, or of missing minutes of audiotape, enigmatic assassination footage, or satellite images of suspected missile sites.

Notes

Introduction

1. See Pingree and Gitelman (2003) for more on this perspective. "Identity crisis" is Altman's (2004) wording, while "transition" is the term favored at MIT and The MIT Press. See Uricchio (2003); Marvin (1988).

2. See Jameson (2003) for a discussion of this point. Jameson wants to leave "aside the question of technological determinism" and yet recognize a symptomatic "projection out of the new media of a whole new set of ideologies appropriate" to the logic of globalized finance capital (705); postmodernism is complete modernism because instantaneous communications have evened out the experience of temporality across the globe as well as between colony and colonizer.

3. On German media studies in general, see Geisler (1999).

4. Here, I am drawing on Latour (1993, 2000). This is Latour's point about anthropology (in this paragraph) and, later, something of his emphasis on the portability of inscriptions. It will be clear, I hope, that I'm not arguing for or against the epistemic conditions I describe (that nature and culture are assiduously kept separate); I am saying that these conditions are vernacular.

5. See de Certeau (1988, 21, 38, passim).

6. On telepresence, see, for instance, Sconce (2000). Admittedly, it has been less common in recent historiography to focus on media as representational than as epistemological, cognitive, and perceptual. Part of the reason for this is again deterministic. Geisler notes how German media studies has no interest in "actual *textuality*" (1999, 79), and explains that this ("programmatic" [88]) avoidance is really an artifact of both technological determinism ("the great advantage of dealing with the media as paradigm-forming technologies is that one need not concern oneself with representation" [104]) and the unadmitted baggage of the high/low assumptions (106–107) that form one legacy of the Frankfurt school.

7. On supervening necessity, see Winston (1998).

8. As McLuhan put it, "It is only too typical that the 'content' of any medium blinds us to the character of the medium" (1964, 9).

9. This comment on the fixity of print is Johns's (1998) point, in a tiny nutshell. On photography, I am drawing on Sandweiss (2002).

10. It's conventional to think of science as bounded and apolitical, but as Shapin and Schaffer (1985, 341–342) show in their account of Robert Boyle, Thomas Hobbes, and the air pump, that conventional boundary itself turns out to have been an artifact of the deeply political socialness from amid which modern science emerged.

11. I have been influenced here by the first chapter of Bowker and Star (1999).

12. Genetics would be different in "the pacing of research and the ways [its] questions may be framed" (Bowker and Star 1999, 36). Similarly, see Lenoir (1994); Clarke and Fujimura (1992); Hankins and Silverman (1995).

13. On the etymology of information, see Nunberg (1996, 111–114). On the reification of information, see Nunberg (1996, 116–123). Hayles writes provocatively about "information losing its body" (1999, 2); for the beginning of this process (the study of information as such), see Day (2000).

14. For the cultural (and rather Geertzian) side of this definition, which moves beyond a more purely semiotic model of communication, see Carey (1989, 13–36). For the technological forms plus protocols side of the definition, I have adapted the discussion of infrastructure in Bowker and Star (1999, chapter 1). The "ontology of representation" points away, I hope, from a purely perceptual account of media forms and toward an unabashedly humanistic sociology of knowledge. I want to get to—if not exactly below—what Mark Hansen calls "the 'threshold' of representation" (2000, 4), because I think that his "holist" approach and the "culturalist" approach can be closer together than he makes out.

15. This is from Lastra (2000, 13), a critique of what he calls "the camera's click" version of media history.

16. Readers may recognize writ large in these questions the determinism of McLuhan-styled media studies (somewhat like Kittler), the Frankfurt school's take on the culture industry (like most of Adorno), and a cultural studies sensibility with which I am admittedly sympathetic, if also leery of its "affirmative" stripe; see Budd, Entman, and Steinman (1990).

17. Bolter and Grusin (1999, 78) make this same observation about their own writing, and I am grateful to Jonathan Auerbach for pointing that out.

18. I have never seen a shred of evidence to suggest that Edison invented the phonograph to save the expense of copper or in thinking of a "Willis-type machine."

19. Geert Lovink, "Interview with Wolfgang Ernst: Archive Rumblings," February 2002, <http://www.perlentaucher.de/buch/10397.html> (accessed May 2005 via <http://laudanum.net/

geert>). I'm drawing on Lovink here because of his superior knowledge of German- and English-language sources.

20. This is Lovink's (2003, 14) "aesthetic undertaking," and Clayton's (2003, 39) "literary" citation of the nineteenth century into postmodernism.

21. For a similarly nuanced periodicity, see Liu (2004b, 63; 2004a).

22. Also helpful here have been Rosenberg (1979), and Manoff's (2004) recent synthesis.

23. Chandler, Davidson, and Johns 2004, 3. I prefer genealogy to archaeology, as it were, or if the label is more Baconian than Foucauldian, this book pursues the "arts of transmission" as recently described in this introduction to a special issue of *Critical Inquiry*.

24. Warner (1993, 9) and Solomon (1993) and Nerone (1993) have also been helpful to my thinking here.

25. See also A. D. Chandler (1997). Both of these accounts tend toward technological determinism, and in my shorthand here, I admittedly run the risk of merely making their cause into my effect and their effect into my cause without sufficiently altering the terms of discussion.

26. Gramophones and phonographs worked in different ways, and could not play the same records.

27. I am thinking particularly of Garvey (1996) and Rabinovitz (1998), but also Miriam Hansen (1991), Rakow (1992), and Fischer (1991, 1992).

28. See Warner (2002) for an insightful elaboration of the Habermasian project. Geisler is describing the work of Helmut Winkler, who has introduced this perspective in German media studies. The centripetal power of media lies behind Anderson's (1991) influential notion of "imagined communities," for one, though again I think there is a danger in conceiving of media as *inherently* centripetal. For a wonderful, teachable example of a community torn apart by reading, see Sarris (1993).

29. There is a growing literature on this first-wave globalization. See, for example, Harold (2001).

30. See Manuel (1993, 37–39); A. Jones (2001, 53); Racy (1977, 97–99); Laird (1999, 18–19); Talking Machine Trade (1911).

31. On Lebanon, see Racy (1977). On Argentina, see Gronow and Saunio (1998, 31).

32. Gronow (1982, 12) estimates that "between 1900 and the 1950s, American companies issued at least 30,000 records aimed at the non-English-speaking communities in the United States," though many titles may only have had issues of one thousand or so.

33. See Farrell (1998, 67), who notes that the regional diversity of the Indian popular music industry was short-lived, and would reappear only with the dissemination of cassette tapes in the 1970s.

34. Recorded sound is easy to overlook as a precedent, I think, because it offers no visual idiom.

35. See, for example, Levinson (1997, xii); Starr (2004, 298–299). The "materiality" of digital media has been the subject of considerable comment among critics—for instance, in Hayles (1999)—though the term itself has developed a range of meanings that make synthesis particularly difficult—an issue I will return to in chapters 3 and 4.

36. As Hayles explained to me in an e-mail in April 2002, "When I coined the phrase 'flickering signifier,' I had in mind a reconfigured relation between the signifier and signified than had been previously articulated in critical and literary theory. As I argue in that piece, the signifier as conceptualized by [Ferdinand de] Saussure and others was conceived as unitary in its composition and flat in its structure. It had no internal structure, whether seen as oral articulation or written mark, that could properly enter into the discourse of semiotics. When signifiers appear on the computer screen, however, they are only the top layer of a complex system of interrelated processes, which MANIFEST themselves as marks to a user, but are properly understood as processes when seen in the context of the digital machine. I hoped to convey this processural quality by the gerund 'flickering,' to distinguish it from the flat durable mark of print or the blast of air associated with oral speech."

37. There is a lot packed into this claim, I know; helpful to this formulation has been J. Chandler's (1998, 60) reflections on (Paul Veyne and) history, as well as Poster's (2001, 73–74) reflections on (Jacques Derrida, Judith Butler, and) the performativity of the trace.

38. "Reading the Background" is the title of a chapter by Brown and Duguid (2000) that elegantly makes this point.

39. A number of scholars have made related assertions. (Reflexivity, as Hayles [1999, 9] observes, is one of the—reflexive—characteristics of contemporary critical theory.) I am thinking of Raymond Williams's (1976) observation about vocabulary in *Keywords,* de Grazia's (1992) observation about the Enlightenment logic of textual authenticity, and J. Chandler's (1998) observation about romantic literary history.

40. Critics who work on film and television make similar observations about "history on" as "history of" with varying degrees of self-consciousness. See, for example, Sobchack (1999–2000); Hanke (2001).

41. Joyce is thinking of hypertext and networked culture generally.

42. See particularly Gillies and Cailliau (2000). Cailliau calls himself (in the third person) "the self-appointed evangelist of the World Wide Web" (324). The Edison/Berners-Lee comparison is from Naughton (2000, 245). The Edison/Moses comparison is from Carlson and Gorman (1990), who call Edison the Moses of mass culture because Moses led the children of Israel to the promised land but did not enter.

43. See Gillies and Cailliau (2000, 218, 226–227); L. Addis, "Brief and Biased History of Preprint and Database Activities at the SLAC Library, 1962–1994," January 2002, <http://

www.slac.stanford.edu/~addis/history.html>. With regard to the classics, I'm thinking of the Perseus project (<http://www.perseus.tufts.edu>) and the *Thesaurus Linguae Graecae;* see Ruhleder (1995).

44. In thinking about disciplines as such, I have been prompted by Ruhleder (1995), and influenced by Lenoir (1997), especially chapter 3.

Chapter 1

1. The portion of this chapter on nickel-in-the-slot phonographs initially was a contribution to a conference at the Dibner Institute at MIT, and I am thankful to Paul Israel and Robert Friedel for their invitation to participate. The material on tinfoil phonographs has occupied me for a long time; different and much more partial versions have appeared as "First Phonographs: Writing and Reading with Sound" and "Souvenir Foils."

2. *Scientific American* 37 (December 1877): 384.

3. Lastra's (2000) first chapter, "Inscriptions and Simulations," is an account of the ways in which modern media were first imagined according to these tropes.

4. There is a widespread misapprehension that magnetic tape was the first medium for amateur recording. Actually, phonographs and graphophones (but not gramophones) could all record sound until electrical recording became the norm around 1920. See Morton (2000).

5. These are claims adapted from the work of Anderson (1991), Habermas (1989), and Warner (1990). On circulation, see also Henkin (1998); John (1995).

6. On public speech, for instance, see Looby (1996); Fliegelman (1993); Grasso (1999); Ruttenburg (1999).

7. The most notable exception is Cmiel (1990).

8. Secord (2000, 523) supposes that "relative stability in print reemerged from the mid- and later 1840s" in Britain, "with the laying of a groundwork for a liberal nation-state, based on imperial free trade and an economic future clearly within the factory system," though this was clearly not the case in the Victorian United States. Another factor unattended here, as in Secord, is electric telegraphy.

9. For the number of papers, see *Centennial Newspaper Exhibition* (1876), where the data were compiled in part from the 1870 census, and in the course of collecting a "monster reading room and an exchange for newspaper men" at the Centennial Exposition in Philadelphia, known as the Newspaper Pavilion, with collected issues from across the United States. According to the *U.S. Census of Manufactures,* in the twenty years between 1880 and 1900, the amount of capital involved in U.S. newspapers and periodicals rose by an estimated 400 percent, and the amount of paper consumed rose by 650 percent.

10. In this last claim, I am agreeing with Secord (2000). Like Secord, I am trying to stress that the fixity of print is a variable social construct rather than an inherent property of the medium, that media are always social before they are perceptual. This case has been stated most strongly by Johns (1998, 2, 19). For the U.S. context, see McGill (2003).

11. See de Grazia (1992, 7), who offers a wonderful account of how the Shakespearean canon emerged as a constituent of this logic, how it "came to be reproduced in a form that continues to be accorded all the incontrovertibility of the obvious."

12. A. Fabian (2000, 173, passim). Slave narratives offer a particularly good example of this.

13. See Gutjahr (1999, 110–111).

14. Secord (2000) makes this point more broadly and at much greater length in his study of the ways that different readers differently received *The Vestiges of the Natural History of Creation* (1844).

15. Article 10 of the *Articles of Impeachment* against Johnson (March 1868) reads, "That said Andrew Johnson . . . did attempt to bring into disgrace, ridicule, hatred, contempt and reproach the Congress of the United States, . . . to excite the odium and resentment of all good people of the United States against Congress . . . and in pursuance of his said design and intent, openly and publicly and before divers assemblages of citizens . . . did . . . make and declare, with a loud voice certain intemperate, inflammatory, and scandalous harangues, and therein utter loud threats and bitter menaces . . . amid the cries, jeers and laughter of the multitudes then assembled in hearing." (The twentieth-century impeachment, by contrast, was less about what President Bill Clinton literally *said* than about what he *meant;* namely, it "depends what the meaning of is is.")

16. In writing about plain speech, Lears relies on Cmiel (1990, 263–265), who counted books of verbal criticism and usage manuals listed in the *National Union Catalog* between S. Hurd's *A Grammatical Corrector* (1847) and R. H. Bell's *The Worth of Words* (1902); Cmiel notes 182 total editions of 34 titles.

17. Emphasis added; quoted in *Clayton v. Stone and Hall,* 2 Payne 392 (1829), referring to "a newspaper or price-current," and cited in *Baker v. Selden,* 101 U.S. 99 (1880). My thinking about the two cases is indebted in part to Meredith L. McGill, "Fugitive Objects: Securing Public Property in United States Copyright Law" (working paper, October 12, 2000).

18. On spirituals, I've been influenced by Cruz (1999).

19. Records of the Edison Speaking Phonograph Company exist at the Edison National Historic Site in West Orange, New Jersey, and at the Historical Society of Pennsylvania in Philadelphia. Documents from West Orange have been microfilmed and form part of the *Thomas A. Edison Papers: A Selective Microfilm Edition* (1987). These items are also available as part of the ongoing electronic edition of the Edison Papers; see <http://edison.rutgers.edu>. For the

items cited here, like Uriah Painter to Thomas Edison, August 2, 1879 ("milked the Exhibition cow"), microfilm reel and frame numbers are given in the following form: TAEM 49:316. Documents from Philadelphia form part of the Painter Papers collection and have been cited as such. The company's incorporation papers are TAEM 51:771. The history of the company may be gleaned from volume 4 of *The Papers of Thomas A. Edison* (1998). Also see Israel (1997–1998).

20. "An Interesting Session Yesterday: Edison, The Modern Magician, Unfolds the Mysteries of the Phonograph," *Washington Star,* April 19, 1878. The *New York Times* carried a shorter version of the same story on the same day.

21. Edward H. Johnson to Uriah H. Painter, January 27, 1878, in the Painter Papers; Johnson prospectus, February 18, 1878, TAEM 97:623. Both are transcribed and published in Edison (1998). I am grateful to the editors of the *Papers of Thomas A. Edison* for sharing this volume in manuscript as well as their knowledge of the Painter Papers.

22. "The Phonograph Exhibited: Prof. Arnold's Description of the Machine in Chickering Hall—Various Experiments, with Remarkable Results," *New York Times,* March 24, 1878, 2:5.

23. Accounts of Jersey City are reported in the *Jersey Journal,* June 13, 14, and 21, 1878, and the *Argus,* June 20, 1878.

24. *Iowa State Register,* July 3, 1878; *Daily Times* (Dubuque), July 16 and June 29, 1878.

25. For a discussion of this improving ethos at later demonstrations, see Musser (1991, chapter 3), where he adapts Neil Harris's idea of an "operational aesthetic" to the circumstances of Lyman Howe's career in "high-class" exhibitions. Differently germane is Cook (2001).

26. "Renewed optimism" is Musser (1991, 22), apropos of Wiebe. In thinking about the demonstrations as hegemonic forms, I have been influenced by Uricchio and Pearson's (1993) work on the "quality" films of 1907–1910; on the pervasiveness and social function of Shakespeare in U.S. culture, see Uricchio and Pearson (1993, 74–78).

27. "Our Washingtons" is from Edison (1878). My carnivalesque touches on Friedrich A. Kittler's association of recorded sound and the Lacanian order of the real, since as Kittler (1999, 16) notes, the phonograph (unlike writing) can record "all the noise produced by the larynx prior to any semiotic order and linguistic meaning."

28. These details from the Painter Papers, Letter books, and Treasurer's books of the Edison Speaking Phonograph Company, including Smith to Hubbard, November 23, 1878; Mason to Redpath, November 1, 1878; Cushing to Redpath, July 16, 1878; Redpath to Mason, July 10, 1878.

29. The year 1878 also saw a bewildering number of versions of *H. M. S. Pinafore* on the U.S. stage, see Allen (1991). An original printed copy of Hockenbery is at the New York Public

Library, available elsewhere in microprint as part of the English and American Drama of the Nineteenth Century series (New York: Redex Microprint, 1968). The production history of *Phunnygraph* is unknown.

30. This refers to the second wave of phonograph demonstrations, when the machine had been "perfected," as Edison put it; *Proceedings of the Fourth Annual Convention of Local Phonograph Companies* (1893), 112. On authorship and exhibitions, see Musser (1991, 6–7).

31. Uriah Painter to Thomas Edison in reference to the Matthew Brady Studio's photograph of the inventor with the phonograph; see TAEM 15:575.

32. We know this figure for Wilde's photographs because they were in violation of copyright; *Burrow-Giles v. Sarony,* 111 U.S. 53 (1884).

33. See Lastra (2000, 16). On sister instruments, see Andem (1892, 7). This point has been difficult for audiences to swallow when I have tried to make it at conferences. I am confident of its accuracy in specific reference to accounts of 1878: though present on occasion, phono/photo analogies are not typical in this first rush of news accounts.

34. From Dickey (1919); noted in Lears (1989, 41), as part of his discussion of advertising and the "modernization of magic."

35. See Menke (2005).

36. "Sacralizing" is from Levine (1988); "salvage" is from James Clifford's "salvage ethnography," noted in Cruz (1999, 180). For the slightly later use of the phonograph in ethnography, see Brady (1999).

37. Stated in the extreme, or too extremely, "All concepts of trace, up to and including Derrida's grammatological ur-writing, are based on Edison's simple idea. The trace preceding all writing, the trace of pure difference still open between reading and writing, is simply a gramophone needle" (Kittler 1999, 33).

38. Quoted and further described in A. Fabian (2000, 132, 134; emphasis added). I am relying on recent scholarship that sees the federal support of Civil War widows, orphans, and veterans as the origin of the U.S. welfare state; see Skocpol (1992).

39. *Weekly* (?), April (?), 1878; see TAEM 25:187. Pat Crain pointed out the "pye" pun to me

40. On Tyler and his style, see Vanderbilt (1986, 81, 84); also see Spengemann (1994, 4–11).

41. *Baker v. Selden* was heard by the U.S. Supreme Court in 1879. In its decision, the Court established the idea/expression dichotomy that has stood in copyright law since. The case involved bookkeeping manuals and asked whether an author's rights to the printed bookkeeping forms extended to cover the bookkeeping method they made possible. The answer was yes at first hearing (in 1874) and no on appeal (101 U.S. 99).

42. An even better example of a similar mistrust of print was the ballad collectors' determination to value unpublished sources above published ones. Francis James Child devoted himself to locating manuscript sources and the sources that "still live on the lips of the people" (Kittredge, 1965, xxvii–xxviii). Childs's work appeared in parts, from 1883 to 1898. American "Folk-Lorists" felt the same way. "Negro" folklorist Alice Mabel Bacon urged, "Nothing must come in that we have ever seen in print" (1897); quoted in Cruz (1999, 170). A generation later, the "new bibliography" movement in literary studies set out to establish editions of canonical "works," salvaged from the (potentially corrupt) printed "texts" of the past.

43. This is Marx (1997, 967) from his work on another keyword.

44. More broadly put, "Production is indeed historiography's quasi-universal principal of explanation, since historical research grasps every document as the symptom of whatever produced it" (de Certeau 1988, 11).

45. The *Oxford English Dictionary* seconds Farmer, noting in 1856 and 1879 the first uses of *record* to mean a person's record, and in 1883 and 1884 the first uses of *record* to mean best recorded achievement.

46. I have been a prime offender here, since I write, infelicitously, that the nickel-in-the-slot phonographs "proved to be a wedge that opened the modern entertainment market" (Gitelman 1999b, 69). This assigns their meaning teleologically, according to what came later, and I here want to ask what they meant in their own time. Kenney's (1999, 23–30) account of the nickel-in-the-slot machines is less teleological. See also Sterne (2003, especially 201–206).

47. My account runs parallel to Kenney's (1999). See also De Graaf (1997–1998); Berliner (1888).

48. This last was reported at the local companies' convention of 1891; see Brooks (1978).

49. "Songs for a Nickel," *Journal* (New York), November 9, 1890.

50. See "The Musical Industry of the Phonograph among Some of Our Companies," *Phonogram* 2 (August–September 1892): 180–188; Andem (1892, 59–61). Sterne's (2003, 137–177) account of "audile technique" is particularly illuminating here (chapter 3). Referring to stethoscopes, phonographs, and telephones, and looking toward radios, Sterne writes, "The technicized, individuated auditory field could be experienced collectively" (161).

51. See also Kasson (1978, 41–50).

52. Quoted in Brooks (1978, 10; emphasis added).

53. "The Musical Industry of the Phonograph," 187; one machine reportedly averaged a take of fifteen dollars a day for three months. That's three hundred nickels for a playing time around two minutes. Six hundred minutes equals ten hours of play per day! Doubtless, the figures were greatly inflated by the *Phonogram,* a trade publication. On the move from parlors to living rooms, see Sterne (2003, 200–209), as well as the chapter that follows below.

54. *Journal* (New York), November 9, 1890; *Times* (Buffalo), May 7, 1892.

55. *Chronicle* (St. Louis), February 14, 1891.

56. *Proceedings of the Second Annual Convention of Local Phonograph Companies* (1891), 64.

57. On fidelity questions and their grounds, see Lastra (2000, chapter 4). See also Thompson (1995); Sterne (2003, chapter 5); Morton (2000, chapter 1). Rick Altman was particularly helpful in the formulation of this paragraph.

58. *Proceedings of the Second Annual Convention of Local Phonograph Companies* (1891), 64.

59. For a description of a rudimentary 1889 catalog, see Brooks (1978, 6). Quoted here is the Columbia catalog dated October 1, 1890, Division of Recorded Sound, Library of Congress, with later catalogs of this and other companies; North American's "Catalogue of Musical Phonograms" is in the archive of the Edison National Historic Site and appears in TAEM 147:361–362. On Columbia and the U.S. Marine Band, see Brooks (1978, 10–18).

60. Edison Phonograph Works's catalog appears as TAEM 147:314–318. For the Louisiana record as well as more on crossover between minstrelsy and recording, see Cogswell (1984, 145).

61. Brooks (1978, 11, 18); Brooks is drawing on research by Ray Wile. For an example of the de facto rationale, see Millard (1995, 80–81); Gaisberg (1942, 83)—hardly an impartial witness given his career in recording—concurs. It's an interesting claim that recurs in the literature, but one that needs more attention.

62. The President's Own, an exhibition by the White House Historical Association, the White House Office of the Curator, and the U.S. Marine Band in cooperation with the National Park Service (permanent).

63. See also Altman (2004, 43–51).

64. Hazen and Hazen (1987, 8).

65. Umble (1996, 63–64, 66) tells the story of two local bands playing through a new telephone line to each other in order to commemorate the event.

66. These generalizations are from Kreitner (1990, 47, 87–88); Kreitner's argument for the typicality of the Pennsylvania county he studies is suggestive without being airtight. On band records, see also Kenney (1999, 28–30).

67. See Ryan (1989).

68. I'm drawing here on Corbin (1998). On civic ceremony in this period, see Ryan (1997, chapter 6).

69. Warner (1990, 61–63); the quote continues, "Printed artifacts were not the only metonym for an abstract public" (62).

70. See Henkin (1998).

71. James Andem, quoted in *Proceedings of the Second Annual Convention of Local Phonograph Companies* (1891), 62–63.

72. Quoted in Brooks (1978, 9, 16).

73. Martland (1997, 16); see also Israel (1997–1998, 41). Sterne (2003, 298) includes only a one-paragraph description of the tinfoil demonstrations, because *The Audible Past* explores neither inscription nor experience as central concerns .

74. Lundy in New Jersey complained bitterly that his territory was always being invaded by others or usurped by Edison's own open-door policy at Menlo Park, while the company for its part suspected that Lundy was underreporting his gross. Redpath's complaint is noted in De Graaf (1997–1998, 48). Lundy's complaint appears in TAEM 19:109, and the company's suspicions were noted by Redpath in TAEM 19:89.

75. Edison Speaking Phonograph, Accounts, TAEM 19:177 and following; for this total, see Israel (1997–1998, 36). For the number of automatic phonographs, see *Proceedings of the Second Annual Convention of Local Phonograph Companies* (1891), 118.

Chapter 2

1. This chapter began long ago as a paper for the Media in Transition conference at MIT (October 1999). That version appears in Thorburn and Jenkins (2003).

2. For telling changes to the meaning of this term (which I am of course applying anachronistically), see Nissenbaum (2004).

3. This is part of the "structural transformation" by which the Habermasian "liberal subject" became personified as (today's denomination) "the consumer." In pointing to this users/publics distinction, I am responding to the present ubiquity of the terms *user* and *user friendly,* and thinking about Siegert's (1998, 79) pronouncement that "the days in which media history is discussed in terms of the 'public sphere' are numbered."

4. See especially chapter 2 of Warner (2002). I am also drawing on Oudshoorn and Pinch's (2003) introduction.

5. This is Purcell's (1995) cogent critique. See also Cockburn (1992); Rosalind Williams (1994).

6. See Fischer (1991). Here, as in his book *America Calling,* Fischer shows how women's uses of the telephone helped to redefine it as an instrument of sociability. Kenney's chapter 5, "The Gendered Phonograph: Women and Recorded Sound, 1890–1930," treats the home phonograph as "a medium for the expression of evolving female gender roles in America" by which women expressed "a range of perspectives, sensibilities, and ambitions that males had not foreseen"

(1999, 89); Kenney's later account of exceptional women indicates those who worked for Victor (an example from 1917) [93], or who staffed the sales counters at department stores in the 1910s and 1920s examples from 1919 and 1925) [95–97]).

7. This is my discomfort with so many design and marketing histories, despite their considerable strengths; for instance, "Although the built environment is designed largely by men, much of it is constructed with female consumers in mind: design thus contributes to the 'making' of modern women" (Lupton 1993, 12). This chapter represents my attempt to push beyond the terms of such accounts using the example of recorded sound.

8. For versions of this narrative, see, for instance, Read and Welch (1976); De Graaf (1997–1998). "Almost by accident" is Starr (2004, 299).

9. No gender is specified for these stenographers; *Proceedings of the First Annual Convention of Local Phonograph Companies,* 57.

10. On feminist histories of technology, I'm thinking gratefully of a panel at the workshop "Science, Medicine, and Technology in the 20th Century: What Difference Has Feminism Made?" Princeton University, October 2–3, 1998. See also (in chronological order) McGaw (1982, 1989); Wajcman (1991); special issue on gender and technology, *Technology and Culture* 38 (January 1997). Nina E. Lerman, Arwen Palmer Mohun, and Ruth Oldenziel, eds. On domestic technology, see also R. Kline (2000, 109, passim).

11. Phonographs are "doubly articulated" as media and technology; see Silverstone and Haddon (1996). "Interpretive flexibility" is a term from the Social Construction of Technology program, outlined by Bijker (1995), among others.

12. According to anthropologist Mary Douglas, "All goods carry meanings, but none by itself. . . . The meaning is in the relations between all the goods." She adds that "goods are used for marking in the sense of clarifying categories" (Douglas and Isherwood 1979, 72, 74).

13. I have adapted this dichotomy of intensive and extensive from both the work of U.S. book historians, where it has been appropriated from Rolf Englesing, and the work of anthropologist Sidney W. Mintz (1985, 152).

14. See Silverstone and Haddon (1996).

15. In thinking about idiosyncratic intensity and fans, I am relying on H. Jenkins (1992, especially 10–45). Barthes is quoted in H. Jenkins (67).

16. The estimate for nickel-in-the-slot era success is from Brooks (1978); the million-selling records appear anecdotally throughout the literature, with the earliest instances either blackface-inspired comedy or Enrico Caruso arias. Million-dollar profits arrived in 1906–1907, judging from the profit and loss sheets of Edison's National Phonograph Company, archive of the Edison National Historic Site (hereafter ENHS).

17. See Garvey (1996); Damon-Moore (1994).

18. On Victor advertising, see Barnum (1991, 29); *Music Trades* 31 (April 7, 1906): 46, cited in Théberge (1997, 102). On Edison, see National Phonograph Company Records, "Advertising" folders, 1906 and other years, ENHS.

19. U.S. Bureau of the Census, *Census of Manufactures 1914* (Washington, DC: Government Printing Office, 1919), 2:825. Notably, print runs for the monthlies were vastly beyond the runs of individual records, which went from several hundred in the late 1890s to several hundred thousand by 1920. Annual record production topped sixty million in the 1920s before plummeting during the Depression.

20. See Strasser (1989).

21. *New York Times,* April 21, 1895, 2:12:1. On Trilby as an exemplary fad, see Abelson (1989, 34). See also E. Purcell (1977); Jenkins (1998); Glenn (2000, 90–91).

22. See Spigel's (1992) authoritative account of television. On radio similarly, see Douglas (1987, chapter 9).

23. *Scientific American* 37 (December 1877): 384. Worth's Palace Museum program from the Theater Collection, New York Public Library. On the business phonograph, see letterhead and advertisements from ENHS. On giddy stenographers, see National Phonograph Company, *The Phonograph and How to Use It* (1900), 140. For the two tropes of inscription and personification as characteristic of modern media, see Lastra (2000, chapter 1).

24. Sterne's (2003, chapter 5; especially 215–225, 274–286) recent account of "The Social Genesis of Sound Fidelity" is particularly helpful in thinking through these complications. Victor and Edison slogans from various advertising copy.

25. Columbia, Boswell, and Bettini catalogs from the collections of the Library of Congress. See also Bottone (1904, 66); Gaisberg (1942, 84).

26. Quoted in Siegert (1998, 87). For a related instance having to do with African American performers as ideal for talking films, see Maurice (2002).

27. Auto Pneumatic Action Company, pictured in Roehl (1973).

28. On stage mimicry, see Glenn (1998, 48–49): "The mimetic moment in American comedy coincided with the mimetic moment in American social thought." The reviewer was Alan Dale in 1908; see Glenn (2000, 84).

29. See Glenn (2000, 84). See also Orvell (1989); Lears (1989).

30. See Roell (1989); *Census of Manufactures 1914,* 2:807–825; Hazen and Hazen (1987); Kreitner (1990).

31. U.S. Bureau of the Census, *Population: Occupational Statistics,* (vol. 4 of *Thirteenth Census of the United States* (Washington, DC: Government Printing Office, 1910); "Music in the American Home," *Good Housekeeping* 39 (1904): 292; James Huneker, "Women and Music," *Harper's Bazar* 33 (1900): 1306–1308, reported in *Current Literature* 39 (1905): 436–437.

32. "Sousa and His Mission," *Music* 16 (1899): 272–276. The following observations by Sousa are from "The Menace of Mechanical Music," *Appleton's* 8 (1906): 278–284. Sousa's band sometimes had a woman harpist for concerts, but was an all male concern.

33. See Roell (1989, 37–45).

34. See Uricchio and Pearson (1993, chapter 3).

35. See Uricchio and Pearson (1993, 67). They suggest that Shakespeare acted as a consensus builder while Dante served as a distinction maker across U.S. culture.

36. In addition to translations among high/low tastes, amateur/professional music, and recorded/live performance, opera records crucially involved the distinction between disc and cylinder records, since the Victor Talking Machine Company so successfully cultivated opera discs as part of its "social reconstruction of the phonograph"; see Kenney (1999, 44–45). Masses and classes are from a 1910 report, National Phonograph Company, "Digest of Conditions of Phonograph Business in New York City and Brooklyn," ENHS.

37. There is no completely satisfactory term for these records, in large measure because of the variability I am addressing here. See Greene (1992); Kenney (1999, chapter 4).

38. In contemplating Sears's pocketknives (131 varieties), Forty (1986, 63, 93) suggests that "to look at the ranges of goods illustrated in the catalogues of nineteenth-century manufacturers, department stores, and mail order houses is to look at a representation of society. . . . One can read the shape of society as manufacturers saw it, and as their customers learned to see it."

39. Spottswood gives all foreign-language records produced in the United States except:

    - Operatic and other classical recording-marketed to all audiences
    - Language instruction records
    - Humorous material employing ethnic stereotypes that was aimed at the general market (for example, the "Cohen" series)
    - Hawaiian music, which (despite language differences) functioned primarily as a variety of U.S. popular music
    - Reissues of non-U.S. matrices on U.S. labels
    - Instrumental recordings by record company "house" bands and orchestras, many of which were issued in the foreign series
    - Native American recordings of private or institutional origin

Despite his commitment to documenting foreign-language records, Spottswood does include Irish and West Indian records in English, which "were treated as foreign-language items by the companies for catalog and distribution purposes" (xvii).

40. See Halttunen (1989); Peiss (1986, 6); Kasson (1978, 41–50).

41. For descriptions and illustrations of the sign, see the in-house *Voice of Victor,* July 1906. For one particularly good reading of the Victor trademark, see Taussig (1993).

42. There is now a rich literature on department stores. "Dream worlds" is Rosalind Williams's (1982) title. See also Abelson (1989); Leach (1993).

43. *Everything Known in Music: A Souvenir of the New Home of the World's Foremost Music House* (Chicago: Lyon and Healy, 1916); available at the New York Public Library. On pluggers and payola, see Segrave (1994).

44. See Joseph McCoy, "Report as to Conditions in the Sale of Edison Phonographs in the State of New York," June 4, 1906, 20; ENHS. There were a few women dealers and a few couples with dealerships; and one notably "up-to-date Jew." The Cobelskill dealer had only six phonographs and four hundred records on hand, while the largest dealers, like two in Utica, had around seventy-five machines and thirty to forty thousand records in stock. Neither the size of the dealership nor the number of dealers in each town was strictly proportional to population. The National Phonograph Company had a total of 8,143 retail dealers in the United States and Canada during the week the report was written; see C. H. Wilson, "Report of Jobbers and Retail Dealers Agreements," June 18, 1906; National Phonograph Company Records, ENHS. Some Edison dealers also handled Columbia goods.

45. See *Talking Machine World* 4, no. 2 (February 1908): 62, the first Sidelines column appeared in *Talking Machine World* 4, no. 1 (January 1908): 67; for later examples, see Kenney (1999, 27). U.S. bike production is estimated by Houndshell (1984, 201) to have been roughly 1.2 million per year during the peak of the craze; see Houndshell (1984, chapter 5). See also Bijker (1995, chapter 2).

46. Steve Wurtzler's forthcoming book has been helpful to this formulation.

47. On the extraterritorial research university, see Leach (1999, 123–149).

Chapter 3

1. *United States v. O'Brien,* 391 U.S. 367 (1968). Congress had overwhelmingly passed an amendment to the Selective Service code in 1965 specifically because it wanted to stop draft-card burning as a form of protest—an argument that O'Brien made fruitlessly to the Court. Flag burning is today's version of this free speech question.

2. Some readers may recognize the pure-text end of things as literary criticism and the untext end of things as museology or material culture studies. Bibliography is the ground in between.

Bookmarks have been examined at length by Peter Stallybrass, and I am grateful to him for sharing his thoughts at several different public presentations that I attended.

3. Bibliographers may blanch at my broad use of *bibliographic,* but I make bold because bibliography forms one of "the most sophisticated branches of media studies we have evolved" (Kirschenbaum 2002, 16). Money admittedly stretches my definition of media as it offers perhaps the hardest case for bibliography.

4. On early efforts to automate banking, see Fischer and McKenney (1993). Vernacular experiences of the materiality of money are particularly hard to trace though important to consider as such. I think, for instance, of the introduction of standard federal currency during the Civil War, late nineteenth-century debates about bimetallism, and the gradual acceptance of credit cards and ATMs starting in the 1970s.

5. See Douglas W. Jones, "Punch Cards: A Brief, Illustrated, Technical History," part of "The Punch Card Collection," updated July 25, 2003, <http://www.cs.uiowa.edu/~jones/cards/> (accessed August 2003). Punch cards were used well before computers, and I've written about these same issues with regard to piano rolls; see Gitelman (2004).

6. *Text* in the current sense of a semiological entity is a concept that emerged in precisely this period, according to Mowitt (1994, 10). In the traditional bibliographic sense, a text is the physical instance of an author's work. Here I mean to evoke both, even though they are somewhat contradictory.

7. See Steve Lubar, "'Do Not Fold, Spindle, or Mutilate': A Cultural History of the Punch Card," May 1991, <http://ccat.sas.upenn.edu/slubar/fsm.html> (accessed August 2003); also in *Journal of American Culture* (Winter 1992).

8. Lubar cites the *Daily Californian,* September 15, 1965, 8, from William Rorabough's study of *Berkeley at War;* see also Draper (1965, 113).

9. Hayles (1999, chapter 4) coins that phrase in her discussion of Norbert Wiener's liberal humanism.

10. Matt Kirschenbaum, "Materiality and Matter and Stuff: What Electronic Texts Are Made Of," October 2001, modified May 2003, <www.electronicbookreview.com>, (accessed June 2003); the same comment appears with elaboration in other works by Kirschenbaum (2000). See also Drucker (2002a). Hayles (1999) offers an account of "how information lost its body," while Drucker, Jerome McGann (2001), and others help to explore the body it still has.

11. Different critics are using the word *materiality* in very different ways. See especially Drucker (1994, 43–46).

12. Kirschenbaum's forthcoming *Mechanisms* (MIT Press) will address the ontological questions at stake, and I'm grateful to Matt for his generous discussions of these and related matters.

13. Sperberg-McQueen (1991) makes a helpful move when he asks, in effect, "What is a representation of a text?" instead of "What is a text?" McGann (2001, especially 91–97) advances on the same turf when he describes writing, rewriting, and re-rewriting the document-type definition for the Rossetti Archive. "Knowledge work" is from Liu (2004a).

14. The phrase is Gladwell's (2002) title, but I am thinking also of Brown and Duguid (2000, especially chapter 7).

15. On porting, see Kirschenbaum (2002, 40–41).

16. Oral histories, interviews, and memoirs remain core sources for many accounts and archives. Manuscript sources are legion and scattered: for an introduction to related records, see Cortada (1990); on Information Processing Techniques Office records in particular, see Norberg and O'Neill (1996, viii–ix). Cortada does not even begin to consider electronic records or archives. On the *New York Times* and the growth of computer science as an academic field, see Rosenzweig (1998); chapter 4 below. For a note on trade literature, see Campbell-Kelley (2003, 25–26).

17. It was more like a grant application process in which grant officers and applicants formed part of the same all-boy, old-boy network, and unofficially, enjoyed hashing out proposals that would later be successfully funded. See Waldrop (2001) for a general account; see Norberg and O'Neill (1996, 54–62) for details. See also Abbate (1999, 54). Even the official request for proposals on the Interface Message Processors was preceded by unofficial conversations among the IPTO and some of the bidders. ARPA didn't actually write its own contracts but had them cut by other government units, usually the Department of Defense.

18. Verner W. Clapp's preface in Licklider (1990, v–ix). Further references to this volume will be indicated by page number in the text.

19. Marvin cites Derek de Solla Price as the best-known proponent of this view.

20. Mumford favored "a reassertion of human selectivity and moral self-discipline, leading to more continent productivity" (182). Like Marvin, Mumford cites Derek Price.

21. Available unpaginated at <www.theatlantic.com>. For the context, see Zachary (1997, 262–269).

22. "Man-Computer Symbiosis" is the title of an important paper Licklider had just published in *IRE Transactions on Human Factors in Engineering* HFE–1 (March 1960): 4–11; available at <http://gatekeeper.dec.com/pub/DEC/SRC/research-reports/abstracts/src-rr-061.html> (accessed August 2003).

23. Parallel to my narrative here would be that of developments at Xerox PARC, where electronic text, networking, and real-time computing were all under development and scrutiny during the same years that the ARPANET got underway. See Levy (2001, chapter 8).

24. Among other reasons for proceeding, "The network itself should be available as a subject for study and experimentation"; Elmer Shapiro, minutes for November 16, 1967 ARAPA Computer Network Working Group meeting at the University of California, Los Angeles, National Archives, Archives II, RG 330 (Department of Defense), 73–A–1647 Box 1.

25. On this principle, see De Rose et al. (1990).

26. Licklider does not use *document* as a verb, also an eighteenth-century usage: "To prove or support [something] by documentary evidence. To provide with documents."

27. See Levy (2001, 21–23).

28. *New York Times,* (March 27, 1966), 171; see also *Washington Post,* (July 7, 1966), H10.

29. See Korte, Myers, and Beery (1960).

30. Poovey works on bookkeeping because she is interested in facticity as an epistemological function, "in how numbers acquired the connotations of transparency and impartiality that have made them so perfectly suited to the epistemological work performed by the modern fact" (5). Mattelart offers an "archaeology of knowledge about communication" that "breaks decisively with the one-side history of systems and theories of communications [to retrace] the genesis of the uses of this term and the multiform realities it has been designating, revealing, or masking" (xv, x).

31. This is an oblique reference to Adrian Johns's *Nature of the Book* (1998) as well as the distinction Johns draws between his work and that of Elizabeth Eisenstein.

32. I am thinking here of two quite different and persuasive histories of history: Grafton (1997) on the footnote, and Batchen (1997) on photography.

33. Crocker's reflections on the birth of the RFC are given in RFC 1000 and RFC 2555; see also "An Interview with Stephen Crocker," Oral History 233, (October 24, 1991), Charles Babbage Institute, Center for the History of Information Processing. See also Kelty (2005). The RFCs themselves are available at <http://rfc-editor.org>, but see below for a note on provenance.

34. Recollection of Jake Feinler in RFC 2555, "Thirty Years of RFCs," April 7, 1999.

35. Doug Engelbart quoted at a NWG meeting in November 1970 by Edwin W. Meyer Jr., as part of RFC 77. On the NIC, see Norberg and O'Neill (1996, 170). To me, one of the most interesting things about the history of the NIC is its role in the Domain Name System adopted in the mid-1980s to structure the Internet. At first, the NIC kept tables of names and computer addresses that it sent out to host computers, but eventually those tables grew so large that they threatened to clog the very network they sought to iterate. The Domain Name System was adopted to solve this problem: that information about a system can overwhelm the very system that it is information about. See Abbate (1999, 189, passim).

36. There were fascinating efforts made periodically to control the accumulated information, like RFC 84 (December 23, 1970), which was rendered obsolete by Karp's RFC 100 (February 26, 1971), itself rendered obsolete by descriptive indexes in subsequent RFCs, like RFC 1000 (August 1987). From what I can tell, these were homegrown efforts that did not enjoy any of the benefits of professional librarianship or bibliography.

37. For a valedictory account of RFCs and Usenet groups, see Hauben and Hauben (1997); see also Abbate (1999, 104, 106–110, 200–201).

38. "Closed world" is from Edwards (1996). Edwards argues that ARPA and its various projects were part of an R & D world that was "closed" by nature of its internal coherence: "Every event was interpreted as part of a titanic struggle between the superpowers" (1; see also 12–22).

39. Jon Postel, RFC 77 (November 20, 1970). This RFC consists of notes taken by Postel and Edwin W. Meyer Jr. at the Houston meetings. Twenty-seven participants took part in one or more of the meetings, including Postel, Meyer, and Engelbart. Karp was the only woman, and at the meetings Crocker "persuaded Peggy Karp to act as NWG/RFC editor. This is a job independent of cataloging RFCs or assigning numbers (functions now performed by NIC). The RFC editor will only categorize RFC as 'hot issues,' current, out of date, or superseded."

40. Postel, RFC 77.

41. See Ceruzzi (2003, 134). See also Wieselman and Tomash (1991).

42. Crocker describes one important way in which differences were portrayed and accommodated, a "protocol fly-off" or programmers' meeting in October 1971 at MIT (RFC 1000); see also Waldrop (2001, 323).

43. Described in Hafner and Lyon (1996, 178–180).

44. Also issued as RFC 254, which is not available online. The sixty-two-page, spiral-bound booklet was produced by the ARPA NIC at SRI. I examined a copy belonging to the University of Washington Libraries.

45. The earliest relevant use of the term *browser* appears to have been in the 1960s, when IBM's Federal Systems Division worked on BROWSER, "An Automatic Indexing On-line Text Retrieval System." This was a recursive acronym, for "BRowsing On-line With SElective Retrieval"; Annual Progress Report by J. H. Williams (September 1969), in Charles Babbage Institute, CBI 32 National Bureau of Standards Computer Literature Collection, Box 465, Folder 4.

46. Oral History, Charles Babbage Institute, quoted in Waldrop (2001, 329).

47. On resource sharing, see Abbate (1999, 104, 106–110).

48. Because it was written and produced at SRI, it might have been composed digitally, using the Online System, but I've seen no evidence of an electronic version.

49. "Scenarios," 19–24.

50. My thinking here has been influenced by Ronald E. Day and Laurent Martinet's partial translation of Suzanne Briet's *Qe'est-ce que la documentation* (Paris: EDIT, 1951), <http://www.lisp.wayne.edu/~ai2398/briet.htm>, (accessed May 2002), especially "Preface to the Translation" and chapter 1.

51. See "History and Philosophy of Project Gutenberg," August 1992, <http://promo.net/pg/history.html> (accessed March 2003). I would like to thank Michael Hart for his personal reflections on the history of Project Gutenberg and its pertinence to my project in this chapter; these reflections were generously offered in extended e-mail exchanges during August 2003. In particular, Hart comments on the politics of his initial publication, "Well, I was VERY rebellious at the time, not that I'm not now, but it wasn't like blowing up the military computer at Michigan or Wisconsin or wherever. We DID force the recruiters out of the Job Fairs. . . . And forced ROTC out of being mandatory around then" (e-mail message, August 13, 2003; ellipsis in original). In 1970, there had been a lot of controversy at Illinois about ARPA's funding for the ILLIAC IV computer, used extensively in Department of Defense research; ILLIAC V was built in California, not Champaign-Urbana.

52. The National Endowment for the Humanities was founded by an act of Congress in 1965, and made its first awards to individuals and institutions (including the CEAA) in 1967. See the National Endowment for the Humanities, "Timeline," <http://www.neh.gov/whoweare/timeline.html> (accessed December 2003).

53. Wilson, who died in 1972, never saw his wish for a U.S. Pléiade realized. The Library of America series was begun in 1979, supported in part by the National Endowment for the Humanities.

54. Mary-Jo Kline (1987, 14–15). Wilson's article was published in September and October installments, and eventually republished as a separate pamphlet, *The Fruits of the MLA* (New York: New York Review Book, 1968) and thence reprinted; see Hancher (1974). The *New York Times* covered the controversy in Benjamin DeMott, "The Battle of the Books," (October 17, 1971), BR58. The entire episode was addressed in Hershel Parker with Bruce Bebb, "The CEAA: An Interim Assessment," *Papers of the Bibliographic Society of America* 68 (1974), 129–148; see also Donald Pizer, "On the Editing of Modern American Texts" and responses, *Bulletin of the New York Public Library* 75 (1971): 147–153.

55. "Almost" because we now pronounce the "dot" in ".com" and the "at" in "@," and so on.

56. It is tempting darkly to discern the origins of the "culture wars" of the 1990s in both the Pentagon-funded ethos of the do-it-ourselves networking community of 1968–1972 and the

Left's own vicious self-criticism in the Fruits of the MLA controversy. I do sense a certain anti-intellectualism in the insular self-sufficiency of the early NWG, particularly in its disregard of professional librarianship. The MLA itself was divided by controversy over the role that its members should or shouldn't play in opposing the Vietnam War; see the *New York Times* editorial, "Breeder of Anti-Intellectualism" (January 1, 1969), denouncing the "antics" of a "noisy fringe group" at the 1969 annual meeting.

57. Quotations are from RFC-Online Project, last modified December 29, 2000, <http://www.rfc-editor.org/rfc-online.html> (accessed November 2003). Current RFC rules are specified in RFC 2223 (October 1997). Lots of RFCs are also available at numerous mirror sites, in Google's Usenet archives, and in guides to the Internet both online and off-line. The republication of RFCs is allowed, though the copyright notice given by the Internet Society urges that any change in format be explained.

58. The Xerox Alto (1973) had a bitmapped screen, so it allowed text editing to be "what you see is what you get", "a phrase made popular by the comedian Flip Wilson on the television program Laugh-In," after he first ad-libbed it in a 1969 episode featuring his cross-dressed alter ego Geraldine Jones (Ceruzzi 2003, 262, 401 n. 53). This etymology for "wissy-wig" is but one suggestion that the phrase "what you see is what you get" cannot be taken literally. A conversation with Paul Ceruzzi in November 2003 was particularly helpful in formulating this then-and-now description, and I am grateful for his observations. On the archival preservation of electronic material, see, for one, the Preservation, Archiving, and Dissemination project of the Electronic Literature Organization, <http://www.eliterature.org/pad/index.asp> (accessed June 2003).

Chapter 4

1. See also Rosenzweig (2004). Roy Rosenzweig confirmed his Lexis/Nexus search in a personal e-mail with the author.

2. ProQuest Company, "ProQuest Historical Newspapers," copyright 2004, < http://www.proquest.com/products/pd-product-HistNews.shtml> (accessed February 2004). Michelle L. Harper of the ProQuest Company was generous in answering my questions about how the digital *Times* was produced.

3. This things-are-people-too sensibility is everywhere in Latour; here I am indebted to Latour (2004). "Dirty ASCII" is the way that ProQuest refers to the initial OCR results, which are then cleaned as employees of the offshore contractor "rekey" headlines and other parts of the text.

4. This is from Seth Lerer's (2002, 17–18) provocative account of errata sheets. The Internet of 1854 is one answer to Lerer's question, "As the age of mechanical reproduction segues into an age of digital transmission, is there any place for error?" (54).

5. Sobchack is writing of QuickTime "movies," but her observations apply in part. ProQuest's *New York Times* is an unaestheticized or a vernacular digital object, and not an object of nostalgia in the way that helps produce Sobchack's "mnemonic aesthetic."
6. See <http://www.w3.org/History>. My use of the verb *invented* is tendentious here, I recognize. For a thorough account of the origins of the World Wide Web, see Gillies and Cailliau (2000); see also Berners-Lee and Fischetti (1999). "A Little History of the World Wide Web" was "created by Robert Cailliau circa 1995," dated by the Web master of the site "Dan Connoly (2000)," and accessed by the author in May 2002 and March 2004.
7. On the contexts of art history lectures, see Nelson (2000). Increasingly, of course, art history students are looking at projections of digital images of works of art.
8. World Wide Web Consortium, "About the World Wide Web," 1992, updated January 24, 2001, <http://www.w3.org/WWW/> (accessed May 2002).
9. See particularly Loizeaux and Fraistat (2002).
10. On "knowledge work," see Liu (2004a, 391–393).
11. "Matter of concern" is Latour again (2004), as is this perspective on documents as explanatory allies. See also Levy (2001, 23) on documents as "talking things" within meaningful contexts—an idea Levy develops in part from Latour.
12. Peter Lyman, "Problem Statement: Why Archive the Web?" Reports and Papers, National Digital Information Infrastructure and Preservation Program, <www.digitalpreservation.gov> (accessed April 2004).
13. See Rosen (1994, 109–117, for example). "Indexical survival" is his terminology.
14. On Benjamin, see Cadava (1997).
15. On television's "overproduction" of history, see Schwoch, White, and Reilly (1992, 3).
16. This last is Gertrude Himmelfarb (1996), quoted in O'Malley and Rosenzweig (1997).
17. See Sobchack (1996, especially 1–14).
18. See Liu (2004b, 8).
19. Price is writing about the novel, not the Web, but her point still stands.
20. John Unsworth, "The Importance of Failure," *Journal of Electronic Publishing* 3 (December 1997), <http://www.press.umich.edu/jep/03-02/unsworth.html> (accessed April 2004).
21. As Liu (2004b, 72) puts it, in the ideological separation of content from presentation means that "knowledge (the great value of postindustrialism) is being separated or extracted from what presentation really means: labor."

22. For example, "These Weapons of Mass Destruction Cannot Be Displayed," <http://www.coxar.pwp.blueyonder.co.uk/> (accessed June 2004). See also "404 Research Lab," <http://www.404lab.com/404/> (accessed June 2004).

23. Forty-four days is Lyman's figure for 2000 and Kahle's for 1997; and more recently, Rick Weiss, "On the Web: Research Work Proves Ephemeral; Electronic Archivists Are Playing Catch-up in Trying to Keep Documents from Landing in History's Dustbin," *Washington Post,* November 24, 2003, A8.

24. Research reported by Scott Carlson, "Here Today, Gone Tomorrow: Studying How Online Footnotes Vanish," *Chronicle of Higher Education,* April 30, 2004, A33. The idea of institutional and disciplinary infrastructure is one being explored by the ACLS Commission of Cyberinfrastructure for the Humanities and Social Sciences, and I am grateful to John Unsworth's introduction, "Do the Humanities Need a Cyberinfrastructure: A Conversation with John Unsworth," at the Washington, DC–Area Forum on Technology and the Humanities, Georgetown University, April 27, 2004. See also Bowker and Star (1999) on infrastructure.

25. On these and related issues concerning the preservation of the Web, see Rosenzweig's (2003) recent call to arms.

26. Both quotes are from the indefatigable Peter Graham, then of Rutgers University Libraries, posting to Newsgroup <bit.listserv.pacs>, October 24, 1994 (accessed via Google Groups, June 2004).

27. As Morris Eaves e-mailed his colleagues at the William Blake Archive project, after looking at listserv postings, "I hadn't realized there were so *many* stabs at standard solutions" (September 11, 1997); hard copy at the Charles Babbage Institute, CBI 174 William Blake Archive, Box 1, Folder 7.

28. Charles Acland (2003, chapter 3) is particularly astute about the temporality of contemporary cinema.

29. For example, Lessig (1999).

30. "Data island" is from Lui (2004b). On the Wayback Machine, see Chris Sullivan, "The Wayback Machine: A Web Archives Search Engine," *Search Day* 127, October 30, 2001, <http://www.searchengine.com> (accessed May 2002).

31. "Remediates" is Bolter and Grusin (1999); this last is a paraphrase of Doane (2002, 24).

32. See also Schwoch, White, and Reilly (1992). Doane is not writing about digital temporality, but her analysis of cinematic time has been helpful to my thinking here.

33. On load times, see Shields (2000, 157–158).

34. Furthermore, remarks Liu (227), "'Real time' online media may not converge on a single, universal intuition of time but instead on multifarious new experiences of time."

35. Marc Andreessen explained of the earliest Mosaic browser (developed at the National Century for Supercomputing Applications at the University of Illinois), its "features" included a "history list per window (both 'where you've been' and 'were you can go')" and a "global history with previously visited locations visually distinct" that is "persistent across sessions"; newsgroup post to <alt.hypertext> and other newsgroups, February 16, 1993, reposted at <www.dejavu.org> by Pär Lannerı (accessed April 2004).

36. Matthew Kirschenbaum, posting to <blake-proj@jefferson.village.virginia.edu>, July 31, 1997; hard copy accessed at the Charles Babbage Institute, CBI 174 William Blake Archive, Box 1, Folder 5.

37. David F. Gallagher, "Don't Mourn, Yet: These Obits Were Only Designs," *New York Times,* (April 21, 2003), C4; CNN closed access to the pages almost immediately, but not before a site called Smoking Gun captured and posted images of them at <www.thesmokinggun.com>.

38. Updated December 6, 2001 (accessed June 2004); see citation in the text below.

39. See also Viscomi (2002); Morris Eaves, "Collaboration Takes More Than E-mail: Behind the Scenes at the William Blake Archive," *Journal of Electronic Publishing* 3, no. 2 (1997), <http://www.press.umich.edu/jep/03-02/blake.html> (accessed May 2004).

40. Dino Buzzetti, "Text Representation and Textual Models," <http://www.iath.virginia.edu/ach-allc.99/proceedings/buzzetti.html> (accessed March 2004).

41. The same phrasing appears in Viscomi (2002).

42. Postings to <blake-proj@jefferson.village.virginia.edu>, October 1997; hard copies accessed at the Charles Babbage Institute, CBI 174 William Blake Archive, Box 1, Folder 6.

43. Google Groups does contain a newsgroup posting by Joseph Viscomi that forwards a copy of the August 4 announcement to <bit.listserv.arils-l> (accessed June 2004).

44. See Wirtén (2004, 134–135).

45. Wirtén (2004, 78).

46. The site <www.dejavu.org> announces its own nostalgia (accessed April 2004). On emulation and preservation strategies, see the overview in Rosenzweig (2003).

47. See R. Johns (1998, 10–11); the annihilation of space was imagined by observers of the U.S. postal system, even before the advent of electric telegraphy.

48. Dan Cohen is developing H-Bot, which as of this writing remains unlinked to the published pages of George Mason's Center for History and New Media, <http://chnm.gmu.edu/tools/

h-bot/> (accessed July 2004). Elena Razlogova introduced me to H-Bot, and I am grateful for her knowledge of it and the other resources available at the center, directed by Roy Rosenzweig. H-Bot is described and quoted by permission of Dan Cohen, July 14, 2004 (boldface in original).

49. For a succinct critique of Google's links-as-votes assumption, see Geoffrey Nunberg, "As Google Goes, So Goes the Nation," *New York Times,* May 18, 2003, Wk4; "When it comes to more specialized topics, the ratings give disproportionate weight to opinions of the activists and enthusiasts that may be at odds with the views of the larger public. It's as if the United Nations General Assembly made all its decisions by referring the question to whichever nation cares most about the issue: the Swiss get to rule on watchmaking, the Japanese on whaling." See also Introna and Nissenbaum (2000).

50. On the Semantic Web ideally, see Lui (2004b); see also Tim Berners-Lee, James Hendler, and Ora Lassila, "The Semantic Web," *Scientific American* (May 2001), <www.sciam.com> (accessed June 2004).

Epilogue

1. *Harper's Weekly,* November 17, 1877, 906; see Fee (1936, 281); "An Indian's View of Indian Affairs," signed "Young Joseph," with an introduction by William H. Hare, missionary bishop of Nebraska, *North American Review* (April 1879): 412–433. See Charles Erskine Scott Wood's "Appendix I" in Fee (330–331). Fee's own version of Joseph is "'From where the sun now stands, I will fight no more against the white man'" (263). Haruo Aoki (1979) analyzes nine different attributed versions in English, starting with one in the *Chicago Times,* October 26, 1877, 120–123. I would like to thank J. Diane Pearson, who directed me to the work of Professor Aoki and described her ongoing project about Niimiipu chief Joseph.

2. This last is to underscore that different technologies have different meanings in different cultural contexts, a point that Miriam Hansen (1999) makes in relation to Chakrabarty's work as well.

3. See Best (2004) for further glosses on *fugitive.* In a new historicist analysis "guided more by analogy than by chronology" (22), Best argues that U.S. intellectual property law draws on elements of the laws regarding chattel slavery and fugitives from slavery.

4. *A Social History of Truth* is the title of a book by Steven Shapin (1994).

5. "Just as the anthropology of religion commences with the explicit or implicit denial of the claims religions make on believers," Gell explains, "so the anthropology of art has to begin with a denial of the claims which objects of art make on the people who live under their spell, and also on ourselves, in so far as we are all self-confessed devotees of the Art Cult" (42). Map the same against the humanities generally: there must be a similarly valuable distinction to be

drawn between studying history while putting aside questions of belief (media history) and studying history while "believing it" (History, with a capital H).

6. Gell critiques the work of Pierre Bourdieu on this ground, because the sociologist "never actually looks at the art object itself, as a concrete product of human ingenuity, but only at its power to mark social distinctions. . . . We have, somehow, to retain the capacity of the aesthetic approach to illuminate the specific objective characteristics of the art object as an object . . . without succumbing to the fascination which all well-made art objects exert on the mind attuned to their aesthetic properties" (42–43).

7. See Chandler, Davidson, and Johns on the arts of transmission (2004, 2).

References

Aarseth, Espen J. 1997. *Cybertext: Perspectives on Ergodic Literature.* Baltimore: Johns Hopkins University Press.

Abbate, Janet. 1999. *Inventing the Internet.* Cambridge: MIT Press.

Abelson, Elaine S. 1989. *When Ladies Go A-Thieving: Middle-Class Shoplifters in the Victorian Department Store.* New York: Oxford University Press.

Acland, Charles R. 2003. *Screen Traffic: Movies, Multiplexes, and Global Culture.* Durham, NC: Duke University Press.

Allen, Robert C. 1991. *Horrible Prettiness: Burlesque and American Culture.* Chapel Hill: University of North Carolina Press.

Altman, Rick. 1984. Toward a Theory of the History of Representational Technologies. *Iris* 2, no. 2:111–125.

Altman, Rick. 2004. *Silent Film Sound.* New York: Columbia University Press.

Andem, James. 1892. *Practical Guide for the Use of the Edison Phonograph.* Cincinnati, OH: C. J. Krehbiel and Co.

Anderson, Benedict. 1991. *Imagined Communities: Reflections on the Origin and Spread of Nationalism.* Rev. ed. London: Verso.

Aoki, Haruo. 1979. *Nez Perce Texts: University of California Publications in Linguistics.* Vol. 90. Berkeley: University of California Press.

Attali, Jacques. 1985. *Noise: The Political Economy of Music.* Trans. Brian Massumi. Minneapolis: University of Minnesota Press.

Barnum, Frederick O., III. *"His Master's Voice" in America.* Camden, NJ: General Electric Company.

Batchen, Geoffrey. 1997. *Burning with Desire: The Conception of Photography.* Cambridge: MIT Press.

Beetham, Margaret. 1990. Towards a Theory of the Periodical as a Publishing Genre. In *Investigating Victorian Journalism,* ed. Laurel Brake, Aled Jones, and Lionel Madden, 19–32. New York: St. Martin's Press.

Beniger, James R. 1986. *The Control Revolution: Technological and Economic Origins of the Information Society.* Cambridge: Harvard University Press.

Benjamin, Walter. 1999. *The Arcades Project.* Trans. Howard Eiland and Kevin McLaughlin, ed. Rolf Teidemann. Cambridge: Harvard University Press.

Berliner, Emile. 1888. The Gramophone: Etching the Human Voice. *Journal of the Franklin Institute* 125 (June): 425–447.

Berners-Lee, Tim, with Mark Fischetti. 1999. *Weaving the Web: The Original Design and Ultimate Destiny of the World Wide Web by Its Inventor.* New York: HarperCollins.

Best, Stephen Michael. 2004. *The Fugitive's Properties: Law and the Poetics of Possession.* Chicago: University of Chicago Press.

Bijker, Wiebe E. 1995. *Of Bicycles, Bakelites, and Bulbs: Toward a Theory of Sociotechnical Change.* Cambridge: MIT Press.

Bogardus, R. F. 1998. The Reorientation of Paradise: Modern Mass Media and Narratives of Desire in the Making of American Consumer Culture. *American Literary History:* 508–523.

Bolter, Jay David, and Richard Grusin. 1999. *Remediation: Understanding New Media.* Cambridge: MIT Press.

Bottone, Selimo Romeo. 1904. *Talking Machines and Records: A Handbook for All Who Use Them.* London: G. Pitman.

Bowker, Geoffrey C., and Susan Leigh Star. 1999. *Sorting Things Out: Classification and Its Consequences.* Cambridge: MIT Press.

Brady, Erika. 1999. *A Spiral Way: How the Phonograph Changed Ethnography.* Jackson: University Press of Mississippi.

Brooks, Tim. 1978. Columbia Records in the 1890s: Founding the Record Industry. *Association for Recorded Sound Collections Journal* 10: 5–36.

Brown, John Seely, and Paul Duguid. 2000. *The Social Life of Information.* Boston: Harvard Business School Press.

Budd, Mike, Robert M. Entman, and Clay Steinman. 1990. The Affirmative Character of U.S. Cultural Studies. *Critical Studies in Mass Communications* 7: 169–184.

Cadava, Eduardo. 1997. *Words of Light: Theses on the Photography of History.* Princeton, NJ: Princeton University Press.

Campbell-Kelley, Martin. 2003. *From Airline Reservation to Sonic the Hedgehog: A History of the Software Industry.* Cambridge: MIT Press.

Cañizares-Esguerra, Jorge. 2001. *How to Write the History of the New World: Histories, Epistemologies, and Identities in the Eighteenth-Century Atlantic World.* Stanford, CA: Stanford University Press.

Carey, James. 1989. *Communication as Culture: Essays on Media and Society.* Boston: Unwin Hyman.

Carlson, W. Bernard, and Michael E. Gorman. 1990. Understanding Invention as a Cognitive Process: The Case of Thomas Edison and Early Motion Pictures, 1888–1891. *Social Studies of Science* 20 (August): 387–430.

Centennial Newspaper Exhibition, 1876. 1876. New York: Geo. P. Rowell and Co.

Ceruzzi, Paul E. 2003. *A History of Modern Computing.* 2nd ed. Cambridge: MIT Press.

Chakrabarty, Dipesh. 2000. *Provincializing Europe: Postcolonial Thought and Historical Difference.* Princeton, NJ: Princeton University Press.

Chandler, Alfred D., Jr. 1977. *The Visible Hand: The Managerial Revolution in American Business.* Cambridge: Harvard University Press.

Chandler, James. 1998. *England in 1819: The Politics of Literary Culture and the Case of Romantic Historicism.* Chicago: University of Chicago Press.

Chandler, James, Arnold I. Davidson, and Adrian Johns. 2004. Arts of Transmission: An Introduction. *Critical Inquiry* 31 (Autumn): 1–6.

Clarke, Adele E., and Joan H. Fujimura, eds. 1992. *The Right Tools for the Job: At Work in Twentieth-Century Life Sciences.* Princeton, NJ: Princeton University Press.

Clayton, Jay. 2003. *Charles Dickens in Cyberspace: The Afterlife of the Nineteenth Century in Postmodern Culture.* New York: Oxford University Press.

Cmiel, Kenneth. 1990. *Democratic Eloquence: The Fight over Popular Speech in Nineteenth-Century America.* Berkeley: University of California Press.

Cockburn, Cynthia. 1992. The Circuit of Technology: Gender, Identity, and Power. In *Consuming Technologies: Media and Information in Domestic Spaces,* ed. Roger Silverstone and Eric Hirsch, 32–47. London: Routledge.

Cogswell, Robert Gireud. 1984. Jokes in Blackface: A Discographic Folklore Study. PhD diss., Indiana University.

Cohen, Lizbeth. 1990. *Making a New Deal: Industrial Workers in Chicago, 1919–1939.* Cambridge: Cambridge University Press.

Cook, James W. 2001. *The Arts of Deception: Playing with Fraud in the Age of Barnum.* Cambridge: Harvard University Press.

Corbin, Alain. 1998. *Village Bells: Sound and Meaning in the 19th-Century French Countryside.* Trans. Martin Thom. New York: Columbia University Press.

Cortada, James W., ed. 1990. *Archives of Data-Processing History: A Guide to Major U.S. Collections.* New York: Greenwood Press.

Cowan, Ruth Schwartz. 1983. *More Work for Mother: The Ironies of Household Technology from the Open Hearth to the Microwave.* New York: Basic Books.

Crary, Jonathan. 1999. *Suspensions of Perception: Attention, Spectacle, and Modern Culture.* Cambridge: MIT Press.

Cruz, Jon. 1999. *Culture on the Margins: The Black Spiritual and the Rise of American Cultural Interpretation.* Princeton, NJ: Princeton University Press.

Curtin, Michael. 2001. Organizing Difference on Global TV: Television History and Cultural Geography. In *Television Histories: Shaping Collective Memory in the Media Age,* ed. Gary R. Edgerton and Peter C. Rollins, 333–356. Lexington: University Press of Kentucky.

Damon-Moore, Helen. 1994. *Magazines for the Millions: Gender and Commerce in the* Ladies' Home Journal *and the* Saturday Evening Post, *1880–1910.* Albany: State University of New York Press.

Daston, Lorraine. 2004. Whither Critical Inquiry? *Critical Inquiry* 30 (Winter): 361–364 .

Day, Ronald E. 2000. The "Conduit Metaphor" and the Nature and Politics of Information Studies. *Journal of the American Society for Information Science* 51: 805–811.

de Certeau, Michel. 1988. *The Writing of History.* Trans. Tom Conley. New York: Columbia University Press.

De Graaf, Leonard. 1997–1998. Thomas Edison and the Origins of the Entertainment Phonograph. *NARAS Journal* 8 (Winter–Spring): 43–69.

de Grazia, Margreta. 1992. *Shakespeare Verbatim: The Reproduction of Authenticity and the 1790 Apparatus.* Oxford: Clarendon Press.

DeLillo, Don. 1985. *White Noise.* New York: Viking.

DeRose, Steven J., David G. Durand, Elli Mylonas, and Allen H. Renear. 1990. What Is Text, Really? *Journal of Computing in Higher Education* 1: 3–26.

Dickey, Marcus. 1919. *The Youth of James Whitcomb Riley: Fortune's Way with the Poet from Infancy to Manhood.* Indianapolis: Bobbs-Merrill.

Doane, Mary Ann. 2002. *The Emergence of Cinematic Time: Modernity, Contingency, the Archive.* Cambridge: Harvard University Press.

Douglas, Mary, and Baron Isherwood. 1979. *The World of Goods.* New York: Basic Books.

Douglas, Susan J. 1987. *Inventing American Broadcasting, 1899–1922.* Baltimore: Johns Hopkins University Press.

Douglass, Frederick. [1892] 1976. *Life and Times of Frederick Douglass, Written by Himself.* New York: Collier Books.

Draper, Hal. 1965. *Berkeley: The New Student Revolt.* New York: Grove Press.

Drucker, Johanna. 1994. *The Visible Word: Experimental Typography and Modern Art, 1909–1923.* Chicago: University of Chicago Press.

Drucker, Johanna. 2002a. Intimations of Immateriality: Graphical Form, Textual Sense, and the Electronic Environment. In *Reimagining Textuality: Textual Studies in the Late Age of Print,* ed. Elizabeth Bergmann Loizeaux and Neil Fraistat, 152–177. Madison: University of Wisconsin Press.

Drucker, Johanna. 2002b. Theory as Praxis: The Poetics of Electronic Textuality. *Modernism/Modernity* 9: 683–691.

Dyer, Richard. 1997. *White.* London: Routledge.

Edison, Thomas. 1878. The Phonograph and Its Future. *North American Review* 126 (June): 527–536.

Edison, Thomas. 1994. *Menlo Park: The Early Years: April 1876–December 1877,* Vol. 3 of *The Papers of Thomas A. Edison,* ed. Robert A. Rosenberg et al. Baltimore: Johns Hopkins University Press.

Edison, Thomas. 1998. *The Wizard of Menlo Park.* Vol. 4 of *The Papers of Thomas A. Edison,* ed. Paul B. Israel, Keith A. Nier, and Louis Carlat. Baltimore: Johns Hopkins University Press.

Edison, Thomas. 1987–. *The Thomas A. Edison Papers: A Selective Microfilm Edition,* ed. Thomas E. Jeffrey et al. Bethesda, MD: University Publications of America.

Edwards, Paul N. 1996. *The Closed World: Computers and the Politics of Discourse in Cold War America.* Cambridge: MIT Press.

Fabian, Ann. 2000. *The Unvarnished Truth: Personal Narratives in Nineteenth-Century America.* Berkeley: University of California Press.

Fabian, Johannes. 1983. *Time and the Other: How Anthropology Makes Its Object.* New York: Columbia University Press.

Farmer, John S. 1889. *Americanisms Old and New.* London: Thomas Poulter.

Farrell, Gerry. 1998. The Early Days of the Gramophone Industry in India: Historical, Social, and Musical Perspectives. In *The Place of Music,* ed. Andrew Leyshon, David Matless, and George Revill, 57–82. New York: Guilford Press.

Fee, Chester Anders. 1936. *Chief Joseph: The Biography of a Great Indian.* New York: Wilson-Erickson.

Fischer, A. W., and J. L. McKenney. 1993. The Development of the ERMA Banking System: Lessons from History. *Annals of the History of Computing, IEEE* 15, no. 1:44–57.

Fischer, Claude S. 1991. "Touch Someone": The Telephone Industry Discovers Sociability. In *Technology and Choice: Readings from Technology and Culture,* ed. Marcel C. LaFollette and Jeffrey K. Stine, 89–116. Chicago: University of Chicago Press.

Fischer, Claude S. 1992. *America Calling: A Social History of the Telephone to 1940.* Berkeley: University of California Press.

Fliegelman, Jay. 1993. *Declaring Independence: Jefferson, Natural Language, and the Culture of Performance.* Stanford, CA: Stanford University Press.

Forty, Adrian. 1986. *Objects of Desire.* New York: Pantheon.

Fukuyama, Francis. 1992. *The End of History and the Last Man.* New York: Free Press.

Gaisberg, F. W. 1942. *The Music Goes Round.* New York: Macmillan.

Garvey, Ellen Gruber. 1996. *Adman in the Parlor: Magazines and the Gendering of Consumer Culture, 1880s–1910s.* New York: Oxford University Press.

Geisler, Michael. 1999. From Building Blocks to Radical Construction: West German Media Theory since 1984. *New German Critique* 78 (Fall): 75–107.

Gell, Alfred. 1992. The Technology of Enchantment and the Enchantment of Technology. In *Anthropology, Art, and Aesthetics,* ed. Jeremy Coote and Anthony Shelton, 40–63. Oxford: Clarendon Press.

Gillespie, Tarleton. 2003. The Stories Digital Tools Tell. In *New Media: Theories and Practices of Digitextuality,* ed. Anna Everett and John T. Caldwell, 107–123. New York: Routledge.

Gillies, James, and Robert Cailliau. 2000. *How the Web Was Born.* Oxford: Oxford University Press.

Ginzburg, Carlo. 2004. Family Resemblances and Family Trees: Two Cognitive Metaphors. *Critical Inquiry* 30: 537–556.

Gitelman, Lisa. 1999a. First Phonographs: Writing and Reading with Sound. *Biblion* 8: 3–16.

Gitelman, Lisa. 1999b. *Scripts, Grooves, and Writing Machines: Representing Technology in the Edison Era.* Stanford, CA: Stanford University Press.

Gitelman, Lisa. 2003. Souvenir Foils. In *New Media, 1740–1915,* ed. Lisa Gitelman and Geoffrey B. Pingree, 157–173. Cambridge: MIT Press.

Gitelman, Lisa. 2004. Media, Materiality, and the Measure of the Digital; or, the Case of Sheet Music and the Problem of Piano Rolls. In *Memory Bytes: History, Technology, and Digital Culture,* ed. Lauren Rabinovitz and Abraham Geil, 199–217. Durham, NC: Duke University Press.

Gladwell, Malcom. 2002. The Social Life of Paper: Looking for Method in the Mess. *New Yorker,* March 25, 92–96.

Glenn, Susan A. 1998. "Give and Imitation of Me": Vaudeville Mimics and the Play of the Self. *American Quarterly* 50 (March): 47–76.

Glenn, Susan A. 2000. *Female Spectacle: The Theatrical Roots of American Feminism*. Cambridge: Harvard University Press.

Graff, Gerald, and Michael Warner. 1989. Introduction: The Origins of Literary Studies in America. In *The Origins of Literary Study in America: A Documentary Anthology*, ed. Gerald Graff and Michael Warner, 1–16. New York: Routledge.

Grafton, Anthony. 1997. *The Footnote: A Curious History*. Cambridge: Harvard University Press.

Grasso, Christopher. 1999. *A Speaking Aristocracy: Transforming Public Discourse in Eighteenth-Century Connecticut*. Chapel Hill: University of North Carolina Press.

Greene, Victor. 1992. *A Passion for Polka: Old-Time Ethnic Music in America*. Berkeley: University of California Press.

Gronow, Pekka. 1982. Ethnic Recordings: An Introduction. In *Ethnic Recordings in America: A Neglected Heritage*. No 1 of *Studies in American Folklife*, American Folklife Center, 1–50. Washington, DC: Library of Congress.

Gronow, Pekka, and Ilpo Saunio. 1998. *An International History of the Recording Industry*. Trans. Christopher Moseley. London: Cassell.

Guillory, John. 1993. *Cultural Capital: The Problem of Literary Canon Formation*. Chicago: University of Chicago Press.

Gutjahr, Paul C. 1999. *An American Bible: A History of the Good Book in the United States, 1777–1880*. Stanford, CA: Stanford University Press.

Habermas, Jürgen. 1989. *The Structural Transformation of the Public Sphere: An Inquiry into a Category of Bourgeois Society*. Trans. Thomas Burger with Frederick Lawrence. Cambridge: MIT Press.

Hafner, Katie, and Matthew Lyon. 1996. *Where Wizards Stay up Late: The Origins of the Internet*. New York: Simon and Schuster.

Halttunen, Karen. 1989. From Parlor to Living Room: Domestic Space, Interior Decoration, and the Culture of Personality. In *Consuming Visions: Accumulation and Display of Goods in America, 1880–1920*, ed. Simon J. Bronner, 157–190. New York: W. W. Norton.

Hancher, Michael. 1974. The Text of "The Fruits of the MLA." *Papers of the Bibliographic Society of America* 68: 411–412.

Hanke, Robert. 2001. Quantum Leap: The Postmodern Challenge of Television as History. In *Television Histories: Shaping Collective Memory in the Media Age*, ed. Gay R. Edgerton and Peter C. Rollins, 59–78. Lexington: University Press of Kentucky.

Hankins, Thomas L., and Robert J. Silverman, eds. 1995. *Instruments and the Imagination*. Princeton, NJ: Princeton University Press.

Hansen, Mark. 2000. *Embodying Technesis: Technology beyond Writing*. Ann Arbor: University of Michigan Press.

Hansen, Miriam. 1991. *Babel and Babylon: Spectatorship in American Silent Film*. Cambridge: Harvard University Press.

Hansen, Miriam Bratu. 1999. The Mass Production of the Senses: Classical Cinema as Vernacular Modernism. *Modernism/Modernity* 6: 59–77.

Harold, James. 2001. *The End of Globalization and the Lessons of the Great Depression*. Cambridge: Harvard University Press.

Hauben, Michael, and Ronda Hauben. 1997. *Netizens: On the History and Impact of Usenet and the Internet*. Los Alamitos, CA: IEEE Computer Society Press.

Hayles, N. Katherine. 1999. *How We Became Posthuman: Virtual Bodies in Cybernetics, Literature, and Informatics*. Chicago: University of Chicago Press.

Hazen, Margaret Hindle, and Robert M. Hazen. 1987. *The Music Men: An Illustrated History of Brass Bands in America, 1800–1920*. Washington, DC: Smithsonian Institution Press.

Henkin, David M. 1998. *City Reading: Written Words and Public Spaces in Antebellum New York*. New York: Columbia University Press.

Himmelfarb, Gertrude. 1996. A Neo-Luddite Reflects on the Internet. *Chronicle of Higher Education*, November 1, A56.

Hockenbery, Frank. 1886. *Prof. Black's Phunnygraph, or Talking Machine: A Colored Burlesque on the Phonograph*. Chicago: T. S. Denison.

Horner, Charles F. 1926. *The Life of James Redpath and the Development of the Modern Lyceum*. New York: Barse and Hopkins.

Houndshell, David A. 1984. *From the American System to Mass Production, 1800–1932: The Development of Manufacturing Technology in the United States*. Baltimore: Johns Hopkins University Press.

Introna, Lucas D., and Helen Nissenbaum. 2000. Shaping the Web: Why the Politics of Search Engines Matters. *Information Society* 16: 169–185.

Israel, Paul. 1997–1998. The Unknown History of the Tinfoil Phonograph. *NARAS Journal* 8 (Winter–Spring): 29–42.

Jameson, Fredric. 2003. The End of Temporality. *Critical Inquiry* 29, 695–718.

Jenkins, Emily. 1998. Trilby: Fads, Photographers, and "Over-Perfect Feet." *Book History* 1: 221–267.

Jenkins, Henry. 1992. *Textual Poachers: Television Fans and Participatory Culture.* New York: Routledge.

John, Richard. 1995. *Spreading the News: The American Postal System from Franklin to Morse.* Cambridge: Harvard University Press.

Johns, Adrian. 1998. *The Nature of the Book: Print and Knowledge in the Making.* Chicago: University of Chicago Press.

Jones, Andrew F. 2001. *Yellow Music: Media Culture and Colonial Modernity in the Chinese Jazz Age.* Durham, NC: Duke University Press.

Jones, Steve, ed. 1999. *Doing Internet Research: Critical Issues and Methods for Examining the Net.* Thousand Oaks, CA: Sage Publications.

Joyce, Michael. 2001. *Othermindedness: The Emergence of Network Culture.* Ann Arbor: University of Michigan Press.

Kamensky, Jane. 1997. *Governing the Tongue: The Politics of Speech in Early New England.* New York: Oxford University Press.

Kasson, John F. 1978. *Amusing the Million: Coney Island at the Turn of the Century.* New York: Hill and Wang.

Kelty, Christopher. 2005. Geeks, Social Imaginaries, and Recursive Publics. *Cultural Anthropology* 20:185–214.

Kenney, William Howland. 1999. *Recorded Music in American Life: The Phonograph and Popular Memory, 1890–1945.* New York: Oxford University Press.

Kirschenbaum, Matthew. 1998. Documenting Digital Images: Textual Meta-Data at the Blake Archive. *Electronic Library* 16 (August), 239–241.

Kirschenbaum, Matthew. 2000. Hypertext. In *Unspun: Key Concepts for Understanding the World Wide Web,* ed. Thomas Swiss, 120–137. New York: New York University Press.

Kirschenbaum, Matthew. 2002. Editing the Interface: Textual Studies and First-Generation Electronic Objects. *Text* 14: 15–50.

Kittler, Friedrich A. 1999. *Gramophone, Film, Typewriter.* Trans. Geoffrey Winthroup-Young and Michael Wutz. Stanford, CA: Stanford University Press.

Kittredge, G. L. 1965. Preface. In *The English and Scottish Popular Ballads,* ed. Francis James Child. 5 vols. New York: Dover.

Kline, Mary-Jo. 1987. *A Guide to Documentary Editing.* Baltimore: Johns Hopkins University Press.

Kline, Ronald R. 2000. *Consumers in the Country: Technology and Social Change in Rural America.* Baltimore: Johns Hopkins University Press.

Korte, Thomas H., Thomas C. Myers, and John W. Beery. 1960. *Microfilm Aperture Card System.* Wright-Patterson Air Force Base, OH: U.S. Air Force.

Kreitner, Kenneth. 1990. *Discoursing Sweet Music: Town Bands and Community Life in Turn-of-the-Century Pennsylvania.* Urbana: University of Illinois Press.

Laird, Ross. 1999. *Sound Beginnings: The Early Record Industry in Australia.* Sydney: Currency Press.

Lastra, James. 2000. *Sound Technology and the American Cinema: Perception, Representation, Modernity.* New York: Columbia University Press.

Latour, Bruno. 1990. Drawing Things Together. In *Representation in Scientific Practice,* ed. Michael Lynch and Steve Woolgar, 19–68. Cambridge: MIT Press.

Latour, Bruno. 1993. *We Have Never Been Modern.* Trans. Catherine Porter. Cambridge: Harvard University Press.

Latour, Bruno. 2000. On the Partial Existence or Existing and Nonexisting Objects. In *Biographies of Scientific Objects,* ed., Lorraine Daston, 247–269. Chicago: University of Chicago Press.

Latour, Bruno. 2004. Why Has Critique Run out of Steam? From Matters of Fact to Matters of Concern. *Critical Inquiry* 30 (Winter): 225–248.

Leach, William. 1993. *Land of Desire: Merchants, Power, and the Rise of a New American Culture.* New York: Vintage.

Leach, William. 1999. *Country of Exiles: The Destruction of Place in American Life.* New York: Vintage Books.

Lears, Jackson. 1989. Beyond Veblen: Rethinking Consumer Culture in America. In *Consuming Visions: Accumulation and Display of Goods in America, 1880–1920,* ed. Simon J. Bronner, 73–98. New York: W. W. Norton.

Lears, Jackson. 1994. *Fables of Abundance: A Cultural History of Advertising in America.* New York: Basic Books.

Lenoir, Timothy. 1994. Helmholtz and the Materialities of Communication. *Osiris* 9: 185–207.

Lenoir, Timothy. 1997. *Instituting Science: The Cultural Production of Scientific Disciplines.* Stanford, CA: Stanford University Press.

Lerer, Seth. 2002. *Error and the Academic Self: The Scholarly Imagination, Medieval to Modern.* New York: Columbia University Press.

Lessig, Lawrence. 1999. *Code and Other Laws of Cyberspace.* New York: Basic Books.

Levine, Lawrence. 1988. *Highbrow/Lowbrow: The Emergence of Cultural Hierarchy in America.* Cambridge: Harvard University Press.

Levinson, Paul. 1997. *The Soft Edge: A Natural History and Future of the Information Revolution*. London: Routledge.

Levy, David M. 2001. *Scrolling Forward: Making Sense of Documents in the Digital Age*. New York: Arcade.

Li, Xia, and Nancy B. Crane. 1996. *Electronic Styles: A Handbook for Citing Electronic Information*. Medford, NJ: Information Today.

Licklider, J. C. R. 1965. *Libraries of the Future*. Cambridge: MIT Press.

Licklider, J. C. R. 1990. The Computer as a Communication Device. In *In Memorium: J. C. R. Licklider*, Palo Alto, CA: Digital Systems Research Center.

Liu, Alan. 2004a. *The Laws of Cool: Knowledge Work and the Culture of Information*. Chicago: University of Chicago Press.

Liu, Alan. 2004b. Transcendental Data: Toward a Cultural History and Aesthetics of the New Encoded Discourse. *Critical Inquiry* 31 (Autumn): 49–84.

Loizeaux, Elizabeth Bergmann, and Neil Fraistat, eds. 2002. *Reimagining Textuality: Textual Studies in the Late Age of Print*. Madison: University of Wisconsin Press.

Looby, Chris. 1996. *Voicing America: Language, Literary Form, and the Origins of the United States*. Chicago: University of Chicago Press.

Lovink, Geert. 2003. *My First Recession: Critical Internet Culture in Transition*. Rotterdam: V_2 Publishing/NAI Publishers.

Lunenfeld, Peter. 1999. Unfinished Business. In *The Digital Dialectic: New Essays on New Media*, ed. Peter Lunenfeld, 7–22. Cambridge: MIT Press.

Lunenfeld, Peter. 2000. *Snap to Grid: A User's Guide to Digital Arts, Media, and Culture*. Cambridge: MIT Press.

Lupton, Ellen. 1993. *Mechanical Brides: Women and Machines from Home to Office*. New York: Cooper-Hewitt National Museum of Design and Princeton Architectural Press.

Manoff, Marlene. 2004. Theories of the Archive from across the Disciplines. *Libraries and the Academy* 4: 9–25.

Manovich, Lev. 2001. *The Language of New Media*. Cambridge: MIT Press.

Manuel, Peter. 1993. *Cassette Culture: Popular Music and Technology in Northern India*. Chicago: University of Chicago Press.

Martland, Peter. 1997. *Since Records Began: EMI, the First 100 Years*. London: B. T. Batsford.

Marvin, Carolyn. 1987. Information and History. In *Ideology of the Information Age*, ed. Jennifer Daryl Slack and Fred Fejes, 49–62. Norwood, NJ: Ablex Publishing.

Marvin, Carolyn. 1988. *When Old Technologies Were New: Thinking about Electric Communication in the Late 19th Century.* New York: Oxford University Press.

Marx, Leo. 1997. Technology: The Emergence of a Hazardous Concept. *Social Research* 64 (Fall): 965–988.

Mattelart, Armand. 1996. *The Invention of Communication.* Trans. Susan Emanuel. Minneapolis: University of Minnesota Press.

Maurice, Alice. 2002. "Cinema at Its Source": Synchronizing Race and Sound in the Early Talkies. *Camera Obscura* 49, no. 17: 31–71.

McGann, Jerome. 1991. *The Textual Condition.* Princeton, NJ: Princeton University Press.

McGann, Jerome. 1996. The Rossetti Archive and Image-Based Electronic Editing. In *The Literary Text in the Digital Age,* ed. Richard J. Finneran, 145–184. Ann Arbor: University of Michigan Press.

McGann, Jerome. 2001. *Radiant Textuality: Literature after the World Wide Web.* New York: Palgrave.

McGaw, Judith A. 1982. Women and the History of American Technology. *Signs* 7: 798–828.

McGaw, Judith A. 1989. No Passive Victims, No Separate Spheres: A Feminist Perspective on Technology's History. In *In Context: History and the History of Technology: Essays in Honor of Melvin Kranzberg,* ed. Stephen H. Cutcliffe and Robert C. Post, 172–191. Bethlehem, PA: Lehigh University Press.

McGill, Meredith L. 2003. *American Literature and the Culture of Reprinting, 1834–1853.* Philadelphia: University of Pennsylvania Press.

McKenney, J. L., and A. W. Fischer. 1993. The Development of the ERMA Banking System: Lessons from History. *Annals of the History of Computing, IEEE* 15, no. 4: 7–26.

McLuhan, Marshall. 1964. *Understanding Media: The Extensions of Man.* New York: McGraw-Hill.

Menand, Louis. 2003. The End Matter: The Nightmare of Citation. *New Yorker,* October 6, 120–126.

Menke, Richard. 2005. Media in America, 1881: Garfield, Guiteau, Bell, Whitman. *Critical Inquiry* 31 (Spring): 638–664.

Millard, Andre. 1995. *America on Record: A History of Recorded Sound.* Cambridge: Cambridge University Press.

Mintz, Sidney W. 1985. *Sweetness and Power: The Place of Sugar in Modern History.* New York: Viking.

Morton, David. 2000. *Off the Record: The Technology and Culture of Sound Recording in America.* New Brunswick, NJ: Rutgers University Press.

Mowitt, John. 1994. *Text: The Genealogy of an Antidisciplinary Object.* Durham, NC: Duke University Press.

Mumford, Lewis. 1968. Emerson behind Barbed Wire. *New York Review of Books* 10, no. 1 (January 18): 3–5.

Mumford, Lewis. 1970. *The Pentagon of Power: The Myth of the Machine.* New York: Harcourt Brace Jovanovich.

Musser, Charles. 1991. *High-Class Moving Pictures: Lyman H. Howe and the Forgotten Era of Traveling Exhibition, 1880–1920.* Princeton, NJ: Princeton University Press.

Naughton, John. 2000. *A Brief History of the Future: From Radio Days to Internet Years in a Lifetime.* Woodstock, NY: Overlook Press.

Nelson, Robert S. 2000. The Slide Lecture, of the Work of Art *History* in the Age of Mechanical Reproduction. *Critical Inquiry* 26 (Spring): 414–434.

Nerone, John C. 1993. A Local History of the Early U.S. Press: Cincinnati, 1793–1848. In *Ruthless Criticism: New Perspectives in U.S. Communication History,* ed. William S. Solomon and Robert W. McChesney, 38–65. Minneapolis: University of Minnesota Press.

Nissenbaum, Helen. 2004. Hackers and the Contested Ontology of Cyberspace. *New Media and Society* 6: 195–217.

Norberg, Arthur L., and Judy E. O'Neill. 1996. *Transforming Computer Technology: Information Processing for the Pentagon, 1962–1986.* Baltimore: Johns Hopkins University Press.

Nunberg, Geoffrey. 1996. Farewell to the Information Age. In *The Future of the Book,* ed. Geoffrey Nunberg, 103–138. Berkeley: University of California Press

Ohmann, Richard. 1996. *Selling Culture: Magazines, Markets, and Class at the Turn of the Twentieth Century.* London: Verso.

O'Malley, Michael, and Roy Rosenzweig. 1997. Brave New World or Blind Alley? American History on the World Wide Web. *Journal of American History* 84: 132–155.

Orvell, Miles. 1989. *The Real Thing: Imitation and Authenticity in American Culture.* Chapel Hill: University of North Carolina Press.

Oudshoorn, Nelly, and Trevor Pinch, eds. 2003. *How Users Matter: The Co-Construction of Users and Technology.* Cambridge: MIT Press.

Peiss, Kathy. 1986. *Cheap Amusements: Working Women and Leisure in Turn-of-the-Century New York.* Philadelphia: Temple University Press.

Pingree, Geoffrey B., and Lisa Gitelman. 2003. Introduction: What's New about New Media. In *New Media, 1740–1915,* ed. Lisa Gitelman and Geoffrey B. Pingree, xi–xxii. Cambridge: MIT Press.

Poovey, Mary. 1998. *A History of the Modern Fact: Problems of Knowledge in the Science of Wealth and Society.* Chicago: University of Chicago Press.

Poster, Mark. 2001. *What's the Matter with the Internet?* Minneapolis: University of Minnesota Press.

Price, Leah. 2000. *The Anthology and the Rise of the Novel.* Cambridge: Cambridge University Press.

Purcell, Carroll. 1995. Seeing the Invisible: New Perceptions in the History of Technology. *Icon* 1: 9–15.

Purcell, Edward L. 1977. Trilby and Trilby-Mania: The Beginning of the Bestseller System. *Journal of Popular Culture* 11: 62–76.

Rabinovitz, Lauren. 1998. *For the Love of Pleasure: Women, Movies, and Culture in Turn-of-the-Century Chicago.* New Brunswick, NJ: Rutgers University Press.

Racy, Ali Jihad. 1977. Musical Change and Commercial Recording in Egypt, 1904–1932. PhD diss., University of Illinois.

Rakow, Lana F. 1992. *Gender on the Line: Women, the Telephone, and Community Life.* Urbana: University of Illinois Press.

Read, Oliver, and Walter L. Welch. 1976. *From Tin Foil to Stereo: Evolution of the Phonograph.* 2nd ed. Indianapolis, IN: Howard W. Sams and Co.

Reiser, Joel Stanley. 1978. *Medicine and the Reign of Technology.* Cambridge: Cambridge University Press.

Roehl, Harvey. 1973. *Player Piano Treasury: The Scrapbook History of the Mechanical Piano in America.* 2nd ed. Vestal, NY: Vestal Press.

Roell, Craig H. 1989. *The Piano in America, 1890–1940.* Chapel Hill: University of North Carolina Press.

Rosen, Philip. 1994. *Change Mummified: Cinema, Historicity, Theory.* Minneapolis: University of Minnesota Press.

Rosenberg, Charles. 1979. An Ecology of Knowledge: On Discipline, Context, and History. In *The Organization of Knowledge in Modern America, 1860–1920,* ed. Alexandra Oleson and John Voss, 440–455. Baltimore: Johns Hopkins University Press.

Rosenzweig, Roy. 1998. Wizards, Bureaucrats, Warriors, and Hackers: Writing the History of the Internet. *American Historical Review* (December): 1530–1547.

Rosenzweig, Roy. 2003. Scarcity or Abundance? Preserving the Past in a Digital Era. *American Historical Review* 108 (June): 735–762.

Rosenzweig, Roy. 2004. How Will the Net's History Be Written? In *The Academy and the Internet,* ed. Helen Nissenbaum and Monroe E. Price, 1–34. New York: Peter Lang.

Ruhleder, Karen. 1995. Reconstructing Artifacts, Reconstructing Work: From Textual Edition to On-Line Databank. *Science, Technology, and Human Values* 20: 39–64.

Ruttenburg, Nancy. 1999. *Democratic Personality: Popular Voice and the Trial of American Authorship.* Stanford, CA: Stanford University Press.

Ryan, Mary P. 1989. The American Parade: Representations of the Nineteenth-Century Social Order. In *The New Cultural History,* ed. Lynn Hunt, 131–153. Berkeley: University of California Press.

Ryan, Mary P. 1997. *Civic Wars: Democracy and Public Life in the American City during the Nineteenth Century.* Berkeley: University of California Press.

Sandweiss, Martha A. 2002. *Print the Legend: Photography and the American West.* New Haven, CT: Yale University Press.

Sarris, Greg. 1993. Keeping Slug Woman Alive: The Challenge of Reading in a Reservation Classroom. In *The Ethnography of Reading,* ed. Jonathan Boyarin, 238–269. Berkeley: University of California Press.

Schreiber, G. R. 1961. *A Concise History of Vending in the U.S.A.* Chicago: Vend, the Magazine of the Vending Industry.

Schwoch, James, Mimi White, and Susan Reilly. 1992. *Media Knowledge: Readings in Popular Culture, Pedagogy, and Critical Citizenship.* Albany: State University of New York Press.

Sconce, Jeffrey. 2000. *Haunted Media: Electronic Presence from Telegraphy to Television.* Durham, NC: Duke University Press.

Sconce, Jeffrey. 2003. Tulip Theory. In *New Media: Theories and Practices of Digitextuality,* ed. Anna Everett and John T. Caldwell, 179–193. New York: Routledge.

Secord, James A. 2000. *Victorian Sensation: The Extraordinary Publication, Reception, and Secret Authorship of Vestiges of the Natural History of Creation.* Chicago: University of Chicago Press.

Segrave, Kerry. 1994. *Payola in the Music Industry: A History, 1880–1991.* Jefferson, NC: McFarland.

Shapin, Steven. 1994. *A Social History of Truth: Civility and Science in Seventeenth-Century England.* Chicago: University of Chicago Press.

Shapin, Steven, and Simon Schaffer. 1985. *Leviathan and the Air-Pump: Hobbes, Boyle, and the Experimental Life.* Princeton, NJ: Princeton University Press.

Shields, Rob. 2000. Hypertext Links: The Ethic of the Index and Its Space-Time Effects. In *The World Wide Web and Contemporary Cultural Theory,* ed. Andrew Herman and Thom Swiss, 144–160. New York: Routledge.

Siegert, Bernard. 1998. Switchboards and Sex: The Nut(t) Case. In *Inscribing Science: Scientific Texts and the Materiality of Communication,* ed. Timothy Lenoir, 78–90. Stanford, CA: Stanford University Press.

Silverstone, Roger, and Leslie Haddon. 1996. Design and the Domestication of Information and Communication Technologies: Technical Change and Everyday Life. In *Communication by Design:*

The Politics of Information and Communication Technologies, ed. Robin Mansell and Roger Silverstone, 44–74. Oxford: Oxford University Press.

Skocpol, Theda. 1992. *Protecting Soldiers and Mothers: The Political Origins of Social Policy in the United States.* Cambridge: Harvard University Press.

Sobchack, Vivian, ed. 1996. *The Persistence of History: Cinema, Television, and the Modern Event.* New York: Routledge.

Sobchack, Vivian. 1999–2000. What Is Film History? Or, the Riddle of the Sphinxes. *Spectator* 20 (Fall–Winter): 8–22.

Sobchack, Vivian. 2004. Nostalgia for a Digital Object: Regrets on the Quickening of QuickTime. In *Memory Bytes: History, Technology, and Digital Culture,* ed. Lauren Rabinovitz and Abraham Geil, 305–329. Durham, NC: Duke University Press.

Solomon, William S. 1993. The Contours of Media History. In *Ruthless Criticism: New Perspectives in U.S. Communication History,* ed. William S. Solomon and Robert W. McChesney, 1–6. Minneapolis: University of Minnesota Press.

Sousa, John Philip. 1906. The Menace of Mechanical Music. *Appleton's Magazine* 8: 278–283.

Spengemann, William C. 1994. *A New World of Words: Redefining Early American Literature.* New Haven, CT: Yale University Press.

Sperberg-McQueen, C. M. 1991. Text in the Electronic Age: Textual Study and Text Encoding, with Examples from Medieval Texts. *Literary and Linguistic Computing* 6: 34–46.

Spigel, Lynn. 1992. *Make Room for TV: Television and the Family Ideal in Postwar America.* Chicago: University of Chicago Press.

Spottswood, Richard K. 1990. *Ethnic Music on Records: A Discography of Ethnic Recordings Produced in the United States, 1893 to 1942.* 6 vols. Urbana: University of Illinois Press.

Starr, Paul. 2004. *The Creation of the Media: Political Origins of Modern Communications.* New York: Basic Books.

Sterne, Jonathan. 2003. *The Audible Past: Cultural Origins of Sound Reproduction.* Durham, NC: Duke University Press.

Stewart, Susan. 1993. *On Longing: Narratives of the Miniature, the Gigantic, the Souvenir, the Collection.* Durham, NC: Duke University Press.

Strasser, Susan. 1989. *Satisfaction Guaranteed: The Making of the American Mass Market.* New York: Pantheon.

The Talking Machine Trade in Japan. 1911. *Talking Machine World* 7, no. 2:19.

Taussig, Michael. 1993. *Mimesis and Alterity: A Particular History of the Senses.* New York: Routledge.

Théberge, Paul. 1997. *Any Sound You Can Imagine: Making Music/Consuming Technology.* Hanover, NH: University Press of New England.

Thompson, Emily. 1995. Machines, Music, and the Quest for Fidelity: Marketing the Edison Phonograph in America, 1877–1925. *Musical Quarterly* 79: 131–171.

Thorburn, David, and Henry Jenkins, eds. 2003. *Rethinking Media Change: The Aesthetics of Transition.* Cambridge: MIT Press.

Tyler, Moses Coit. 1878. *A History of American Literature, 1607–1765.* New York: Putnam's.

Umble, Diane Zimmerman. 1996. *Holding the Line: The Telephone in Old Order Mennonite and Amish Life.* Baltimore: Johns Hopkins University Press.

Uricchio, William. 2003. Historicizing Media in Transition. In *Rethinking Media Change: The Aesthetics of Transition,* ed. David Thorburn and Henry Jenkins, 23–38. Cambridge: MIT Press.

Uricchio, William, and Roberta E. Pearson. 1993. *Reframing Culture: The Case of the Vitagraph Quality Films.* Princeton, NJ: Princeton University Press.

Vanderbilt, Kermit. 1986. *American Literature and the Academy: The Roots, Growth, and Maturity of a Profession.* Philadelphia: University of Pennsylvania Press.

Veysey, Lawrence. 1979. The Plural Organized Worlds of the Humanities. In *The Organization of Knowledge in Modern America, 1860–1920,* ed. Alexandra Oleson and John Voss, 51–106. Baltimore: Johns Hopkins University Press.

Viscomi, Joseph. 2002. Digital Facsimiles: Reading the William Blake Archive. *Computers and the Humanities* 36: 27–48.

Wajcman, Judy. 1991. *Feminism Confronts Technology.* University Park: Penn State University Press.

Waldrop, M. Mitchell. 2001. *The Dream Machine: J. C. R. Licklider and the Revolution That Made Computing Personal.* New York: Penguin Books.

Warner, Michael. 1990. *The Letters of the Republic: Publication and the Public Sphere in Eighteenth-Century America.* Cambridge: Harvard University Press.

Warner, Michael. 1993. The Public Sphere and the Cultural Mediation of Print. In *Ruthless Criticism: New Perspectives in U.S. Communication History,* ed. William S. Solomon and Robert W. McChesney, 7–37. Minneapolis: University of Minnesota Press.

Warner, Michael. 2002. *Publics and Counterpublics.* New York: Zone Books.

Weinberger, David. 2002. *Small Pieces Loosely Joined: A Unified Theory of the Web.* New York: Perseus Books.

Wiebe, Robert H. 1967. *The Search for Order: 1877–1920.* New York: Hill and Wang.

Wieselman, Irving L., and Erwin Tomash. 1991. Marks on Paper: Part I. A Historical Survey of Output Printing. *IEEE Annals of the History of Computing* 13: 63–79.

Williams, Mark. 2003. Real-Time Fairly Tales: Cinema Prefiguring Digital Anxiety. In *New Media: Theories and Practices of Digitextuality,* ed. Anna Everett and John T. Caldwell, 159–178. New York: Routledge.

Williams, Raymond. [1974] 1992. *Television: Technology and Cultural Form.* Hanover, NH: Wesleyan University Press.

Williams, Raymond. 1976. *Keywords: A Vocabulary of Culture and Society.* New York: Oxford University Press.

Williams, Rosalind H. 1982. *Dream Worlds: Mass Consumption in Late Nineteenth-Century France.* Berkeley: University of California Press.

Williams, Rosalind H. 1994. The Political and Feminist Dimensions of Technological Determinism. In *Does Technology Drive History? The Dilemma of Technological Determinism,* ed. Merritt Roe Smith and Leo Marx, 217–235. Cambridge: MIT Press.

Wilson, Edmund. 1968. *The Fruits of the MLA.* New York: New York Review of Books.

Winston, Brian. 1998. *Media Technology and Society: A History from the Telegraph to the Internet.* London: Routledge.

Winthrop-Young, Geoffrey, and Michael Wutz. 1999. Translator's Introduction: Friedrich Kittler and Media Discourse Analysis. In *Gramophone, Film, Typewriter,* by Friederich A. Kittler, trans. Geoffrey Winthrop-Young and Michael Wutz, xi–xxxviii. Stanford, CA: Stanford University Press.

Wired. 2002. Special history issue. 10.01 (January).

Wirtén, Eva Hemmungs. 2004. *No Trespassing: Authorship, Intellectual Property Rights, and the Boundaries of Globalization.* Toronto: University of Toronto Press.

Zachary, G. Pascal. 1997. *Endless Frontier: Vannevar Bush, Engineer of the American Century.* New York: Free Press.

Index